T0296249

CAMBRIDGE LIBRARY COLLECTION

Books of enduring scholarly value

Physical Sciences

From ancient times, humans have tried to understand the workings of the world around them. The roots of modern physical science go back to the very earliest mechanical devices such as levers and rollers, the mixing of paints and dyes, and the importance of the heavenly bodies in early religious observance and navigation. The physical sciences as we know them today began to emerge as independent academic subjects during the early modern period, in the work of Newton and other 'natural philosophers', and numerous sub-disciplines developed during the centuries that followed. This part of the Cambridge Library Collection is devoted to landmark publications in this area which will be of interest to historians of science concerned with individual scientists, particular discoveries, and advances in scientific method, or with the establishment and development of scientific institutions around the world.

Popular Lectures and Addresses

William Thomson, Baron Kelvin (1824–1907), was educated at Glasgow and Cambridge. While only in his twenties, he was awarded the University of Glasgow's chair in natural philosophy, which he was to hold for over fifty years. He is best known through the Kelvin, the unit of measurement of temperature named after him in consequence of his development of an absolute scale of temperature. These volumes collect together Kelvin's lectures for a wider audience. In a convivial but never condescending style, he outlines a range of scientific subjects to audiences of his fellow scientists. The range of topics covered reflects Kelvin's broad interests and his stature as one of the most eminent of Victorian scientists. Volume 1, published in 1889, includes talks about the constitution of matter and basic topics in physics such as light, heat, electricity and gravity.

Cambridge University Press has long been a pioneer in the reissuing of out-of-print titles from its own backlist, producing digital reprints of books that are still sought after by scholars and students but could not be reprinted economically using traditional technology. The Cambridge Library Collection extends this activity to a wider range of books which are still of importance to researchers and professionals, either for the source material they contain, or as landmarks in the history of their academic discipline.

Drawing from the world-renowned collections in the Cambridge University Library, and guided by the advice of experts in each subject area, Cambridge University Press is using state-of-the-art scanning machines in its own Printing House to capture the content of each book selected for inclusion. The files are processed to give a consistently clear, crisp image, and the books finished to the high quality standard for which the Press is recognised around the world. The latest print-on-demand technology ensures that the books will remain available indefinitely, and that orders for single or multiple copies can quickly be supplied.

The Cambridge Library Collection will bring back to life books of enduring scholarly value (including out-of-copyright works originally issued by other publishers) across a wide range of disciplines in the humanities and social sciences and in science and technology.

Popular Lectures and Addresses

VOLUME 1:
CONSTITUTION OF MATTER

LORD KELVIN

CAMBRIDGE
UNIVERSITY PRESS

CAMBRIDGE UNIVERSITY PRESS

Cambridge, New York, Melbourne, Madrid, Cape Town,
Singapore, São Paolo, Delhi, Tokyo, Mexico City

Published in the United States of America by Cambridge University Press, New York

www.cambridge.org
Information on this title: www.cambridge.org/9781108029773

© in this compilation Cambridge University Press 2011

This edition first published 1889
This digitally printed version 2011

ISBN 978-1-108-02977-3 Paperback

POPULAR LECTURES

AND

ADDRESSES.

VOL. I.   *a*

NATURE SERIES.

POPULAR LECTURES

AND

ADDRESSES

BY

SIR WILLIAM THOMSON, LL.D., F.R.S., F.R.S.E., &c.

PROFESSOR OF NATURAL PHILOSOPHY IN THE UNIVERSITY OF GLASGOW, AND
FELLOW OF ST. PETER'S COLLEGE, CAMBRIDGE.

IN THREE VOLUMES

VOL. I.

CONSTITUTION OF MATTER

WITH ILLUSTRATIONS

London:

MACMILLAN AND CO.

AND NEW YORK

1889

RICHARD CLAY AND SONS, LIMITED,
LONDON AND BUNGAY.

PREFACE.

SHORTLY after the delivery of my lecture "On Capillarity" at the Royal Institution, in January 1886, it was suggested to me by Mr. Lockyer, that it might be advisable to make that lecture more easily and more conveniently accessible than it could be in the "Transactions of the Royal Institution" or in the pages of *Nature*. It was accordingly arranged to bring out, as one of the *Nature* series, a small volume containing the lecture "On Capillarity," together with several other papers pertinent to the subject.

While the earlier sheets of this book were passing

through the printer's hands, it occurred to me that it might be well to reissue in a collected form several other lectures and addresses of a popular character, which I have given from time to time, and which could not find a fitting place in my "Reprint of Mathematical and Physical Papers," now being published by the Cambridge University Press. After consideration it was decided to change the character of the proposed volume "On Capillarity," and to increase its size and make it the first of a series of three volumes to constitute a reprint of all my popular lectures and addresses.

The order in which the various articles are arranged, both in the present volume and in those which are to follow, is, generally speaking, according to the subject matter. Thus in the present volume are included lectures concerned with the ultimate constitution of matter. The second volume will include subjects connected with geology, and the third will be chiefly concerned with phenomena of the ocean and maritime affairs.

The lectures are reprinted practically in the form in which they originally appeared, the only alterations being slight verbal changes introduced in a few cases solely for the sake of clearness.

WILLIAM THOMSON.

THE UNIVERSITY, GLASGOW.
Dec. 21, 1888.

CONTENTS.

b

Popular Lectures and Addresses.

CAPILLARY ATTRACTION.

Friday Evening Lecture before the Royal Institution, January 29, 1886 (*Proc. Roy. Inst.*, vol. xi. part III)].

THE heaviness of matter had been known for as many thousand years as men and philosophers had lived on the earth, but none had suspected or imagined, before Newton's discovery of universal gravitation, that heaviness is due to action at a distance between two portions of matter. Electrical attractions and repulsions, and magnetic attractions and repulsions, had been familiar to naturalists and philosophers for two or three thousand years. Gilbert, by showing that the earth, acting as a great magnet, is the efficient cause of the compass needle's pointing to the north, had enlarged people's ideas regarding the distances at which magnets can exert sensible

B

action. But neither he nor any one else had suggested that heaviness is the resultant of mutual attractions between all parts of the heavy body and all parts of the earth, and it had not entered the imagination of man to conceive that different portions of matter at the earth's surface, or even the more dignified masses called the heavenly bodies, mutually attract one another. Newton did not himself give any observational or experimental proof of the mutual attraction between any two bodies, of which both are smaller than the moon. The smallest case of gravitational action which was included in the observational foundation of his theory, was that of the moon on the waters of the ocean, by which the tides are produced ; but his inductive conclusion that the heaviness of a piece of matter at the earth's surface, is the resultant of attractions from all parts of the earth acting in inverse proportion to squares of distances, made it highly probable that pieces of matter within a few feet or a few inches apart attract one another according to the same law of distance, and Cavendish's splendid experiment verified this conclusion. But now for our question of this evening. Does this attraction between any particle of matter in one body and any particle of matter in another continue to vary inversely as the square of the distance, when the distance between the nearest points of the two bodies is diminished to an inch (Cavendish's experiment does not demonstrate this, but makes it very probable), or to a centimetre, or to the hundred-thousandth of a centimetre, or to the hundred-millionth of a centimetre? Now I dip my finger into this basin of water; you see proved a force of attraction between the finger and the drop hanging from it, and between the matter on the two sides of any horizontal plane you like to imagine through the hanging water. These forces are millions of times greater than what you would calculate from the Newtonian law, on the supposition that water is perfectly homogeneous. Hence either these forces of attraction must, at very small distances, increase enormously more rapidly than according to the Newtonian law, or the substance of water is not homogeneous. We now know that it is not homogeneous. The Newtonian theory of gravitation is not surer to us now than is the atomic or molecular theory in chemistry and physics; so far, at all events, as its assertion of heterogeneousness in the minute structure of matter, apparently homogeneous to our senses and to our most delicate direct instrumental tests. Hence, unless we find heterogeneousness and the Newtonian law of attraction incapable of explaining cohesion and capillary attraction, we are not forced to seek the explanation in a deviation from Newton's law of gravitational force. In a communication to the Royal Society of Edinburgh twenty-four years ago,[1] I showed that heterogeneousness does suffice to account for any force of cohesion, however great, provided only we give sufficiently great density to the molecules in the heterogeneous structure.

Nothing satisfactory, however, or very interesting mechanically, seems attainable by any attempt to

---

[1] "Note on Gravity and Cohesion," *Proceedings of the Royal Society of Edinburgh*, April 21, 1862 (vol. iv.). This paper is reprinted in full as Appendix B to the present article.

work out this theory without taking into account the molecular motions which we know to be inherent in matter, and to constitute its heat. But so far as the main phenomena of capillary attraction are concerned, it is satisfactory to know that the complete molecular theory could not but lead to the same resultant action in the aggregate as if water and the solids touching it were each utterly homogeneous to infinite minuteness, and were acted on by mutual forces of attraction sufficiently strong between portions of matter which are exceedingly near one another, but utterly insensible between portions of matter at sensible distances. This idea of attraction insensible at sensible distances (whatever molecular view we may learn, or people not now born may learn after us, to account for the innate nature of the action,) is indeed the key to the theory of capillary attraction, and it is to Hawkesbee [1] that we owe it. Laplace [2] took it up

[1] *Transactions Royal Society*, vols. xxvi., xxvii. 1709—1713 ; or abridged edition, by Dr. Hutton and others, vol. v. p. 464, *et seq.*

[2] *Mécanique Céleste*, supplement to the tenth book, published 1806 ; also *Supplément à la Théorie de l'Action capillaire*, forming a second supplement to the tenth book.

and thoroughly worked it out mathematically in a very admirable manner.   One part of the theory which he left defective—the action of a solid upon a liquid, and the mutual action between two liquids —was made dynamically perfect by Gauss,[1] and the finishing touch to the mathematical theory was given by Neumann[2] in stating for liquids the rule corresponding to Gauss's rule for angles of contact between liquids and solids.

Gauss, expressing enthusiastic appreciation of Laplace's work, adopts the same fundamental assumption of attraction sensible only at insensible distances, and, while proposing as chief object to complete the part of the theory not worked out by his predecessor, treats the dynamical problem afresh in a remarkably improved manner, by founding it wholly upon the principle of what we now call potential energy.   Thus, though the formulas in which he expresses mathematically

---

[1] *Principia generalia Theoriæ Figuræ Fluidorum in Statu Equilibrii* (Göttingen, 1830) ; or *Werke*, vol. v. 29 (Göttingen, 1887).

Herr F. E. Neumann.

his ideas are scarcely less alarming in appearance than those of Laplace, it is very easy to translate them into words by which the whole theory will be made perfectly intelligible to persons who imagine themselves incapable of understanding sextuple integrals. Let us place ourselves conveniently at the centre of the earth so as not to be disturbed by gravity. Take now two portions of water, and let them be shaped over a certain area of each—call it A for the one and B for the other—so that when put together they will fit perfectly throughout these areas. To save all trouble in manipulating the supposed pieces of water, let them become for a time perfectly rigid, without, however, any change in their mutual attraction. Bring them now together till the two surfaces A and B come to be within the one-hundred-thousandth of an inch apart, that is, the forty-thousandth of a centimetre, or two hundred and fifty micro-millimetres (about half the wavelength of green light). At so great a distance the attraction is quite insensible: we may feel very confident that it differs, by but a small percentage, from the exceedingly small force of attraction which we should calculate for it according to the Newtonian law, on the supposition of perfect uniformity of density in each of the attracting bodies. Well known phenomena of bubbles, and of watery films wetting solids, make it quite certain that the molecular attraction does not become sensible until the distance is much less than 250 micro-millimetres. From the consideration of such phenomena Quincke[1] came to the conclusion that the molecular attraction does become sensible at distances of about fifty micro-millimetres. His conclusion is strikingly confirmed by the very important discovery of Reinold and Rücker[2] that the black film, always formed before an undisturbed soap bubble breaks, has a uniform or nearly uniform thickness of about eleven or twelve micro-millimetres. The abrupt commencement and the permanent stability of the black film, demonstrate a proposition of fundamental im-

[1] *Pogg. Ann. der Phys. und der Chem.* Bd. cxxxvii. 1869.

[2] *Proc. Roy. Soc.* June 21, 1877 ; and *Trans. Roy. Soc.* April 19, 1883.

portance in the molecular theory :—The tension of the film, which is sensibly constant when the thickness exceeds fifty micro-millimetres, diminishes to a minimum, and begins to increase again when the thickness is diminished to ten micro-millimetres. It seems not possible to explain this fact by any imaginable law of force between the different portions of the film supposed homogeneous, and we are forced to the conclusion that it depends upon molecular heterogeneousness. When the homogeneous molar theory is thus disproved by observation, and its assumption of a law of attraction augmenting more rapidly than according to the Newtonian law when the distance becomes less than fifty micro-millimetres is proved to be insufficient, may we not go farther and say that it is unnecessary to assume any deviation from the Newtonian law of force varying inversely as the square of the distance, continuously from the millionth of a micro-millimetre to the distance of the remotest star or remotest piece of matter in the universe ; and, until we see how gravity itself is to be explained, as Newton and Faraday thought it must be explained, by some continuous action of intervening or surrounding matter, may we not be temporarily satisfied to explain capillary attraction merely as Newtonian attraction intensified in virtue of intensely dense molecules movable among one another, of which the aggregate constitutes a mass of liquid or solid.

But now for the present, and for the rest of this evening, let us dismiss all idea of molecular theory, and think of the molar theory pure and simple, of Laplace and Gauss ; returning to our two pieces of rigidified water left at a distance of 250 micro-millimetres from one another. Holding them in my two hands, I let them come nearer and nearer until they touch all along the surfaces A and B. They begin to attract one another with a force which may be scarcely sensible to my hands when their distance apart is fifty micro-millimetres, or even as little as ten micro-millimetres ; but which certainly becomes sensible when the distance becomes one micro-millimetre, or the fraction of a micro-millimetre ; and enormous, hundreds or thousands of kilogrammes weight, before they come into absolute contact. I am supposing the area of each of the opposed surfaces to be a few square centimetres. To fix the ideas, I shall suppose it to be exactly thirty square centimetres. If my sense of force were sufficiently metrical, I should find that the work done by the attraction of the rigidified pieces of water in pulling my two hands together, was just about four and a half centimetre-grammes. The force to do this work, if it had been uniform throughout the space of fifty micro-millimetres (five-millionths of a centimetre) must have been nine hundred thousand grammes weight, that is to say, nine-tenths of a ton. But in reality it is done by a force increasing from something very small at the distance of fifty micro-millimetres to some unknown greatest amount. It may reach a maximum before absolute contact, and then begin to diminish, or it may increase and increase up to contact, we cannot tell which. Whatever may be the law of variation of the force, it is certain that throughout a small part of the distance it is considerably more than one ton. It is possible that it is enormously more than one ton, to make up the ascertained amount of work of four and a half centimetre-grammes performed in a space of fifty micro-millimetres.

But now let us vary the circumstances a little. I take the two pieces of rigidified water, and bring them to touch at a pair of corresponding points in the borders of the two surfaces A and B, keeping the rest of these surfaces wide asunder (see Fig. 1). The work done on my hands in this proceeding is infinitesimal. Now, without at all altering the law of attractive force, let a minute film of the rigidified water become fluid all over each of the surfaces A and B: you see exactly what takes place. The pieces of matter I hold in my hands are not the supposed pieces of rigidified water. They are glass, with the surfaces A and B thoroughly cleaned and wetted all over each with a thin film of water. What you now see taking place is the same as what would take place if things were exactly according to our ideal supposition. Imagine, therefore, that these are

really two pieces of water, all rigid, except the
thin film on each of the surfaces A and B, which
are to be put together. Remember also that the

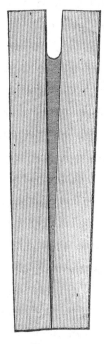

Fig. 1.

Royal Institution, in which we are met, has been
for the occasion transported to the centre of the
earth, so that we are not troubled in any way by

gravity. You see we are not troubled by any
trickling down of these liquid films—-but I must
not say *down*, we have no up and down here—they
do not trickle along these surfaces towards the
table, at least you must imagine that they do not
do so. I now turn one or both of these pieces
of matter till they are so nearly in contact all
over the surfaces A and B, that the whole inter-
stice becomes filled with water. My metrical
sense of touch tells me that exactly four and a
half centimetre-grammes of work has again been
done ; this time, however, not by a very great
force through a space of less than fifty micro-
millimetres, but by a very gentle force acting
throughout the large space of the turning or fold-
ing-together motion which you have seen, and
now see again. We know, in fact, by the elementary principle of work done in a conservative
system, that the work done in the first case of
letting the two bodies come together directly,
and in the second case of letting them come
together by first bringing two points into contact
and then folding them together, must be the

same, and my metrical sense of touch has merely told me, in this particular sense, what we all know theoretically must be true in every case of proceeding by different ways to the same end from the same beginning.

Now in this second way we have, in performing the folding motion, allowed the water surface to become less by sixty square centimetres. It is easily seen that, provided the radius of curvature in every part of the surface exceeds one or two hundred times the extent of distance to which the molecular attraction is sensible, or, as we may say practically, provided the radius of curvature is everywhere greater than 5,000 micro-millimetres (that is, the two-hundredth of a millimetre), we should have obtained this amount of work with the same diminution of water-surface, however performed. Hence our result is that we have found 4·5/60 (or 3/40) of a centimetre-gramme of work per square centimetre of diminution of surface. This is precisely the result we should have had if the water had been absolutely deprived of the attractive force between water and water, and its

whole surface had been coated over with an infinitely thin contractile film possessing a uniform
contractile force of 3/40 of a gramme weight, or
seventy-five milligrammes, per lineal centimetre.

It is now convenient to keep to our ideal film,
and give up thinking of what, according to our
present capacity for imagining molecular action,
is the more real thing—namely, the mutual attraction between the different portions of the liquid.
But do not, I entreat you, fall into the paradoxical
habit of thinking of the surface film as other than
an ideal way of stating the resultant effect of
mutual attraction between the different portions
of the fluid. Look, now, at one of the pieces of
water ideally rigidified, or if you please, at the
two pieces put together to make one. Remember
we are at the centre of the earth. What will
take place if this piece of matter resting in the
air before you suddenly ceases to be rigid? Imagine it, as I have said, to be enclosed in a film
everywhere tending to contract with a force equal
to 3/40 of a gramme or seventy-five milligrammes
weight per lineal centimetre. This contractile film

will clearly press most where the convexity is greatest. A very elementary piece of mathematics tells us that on the rigid convex surface which you see, the amount of its pressure per square centimetre will be found by multiplying the sum [1] of the curvatures in two mutually-perpendicular normal sections, by the amount of the force per lineal centimetre. In any place where the surface is concave the effect of the surface tension is to suck outwards—that is to say, in mathematical language, to exert negative pressure inwards. Now, suppose in an instant the rigidity to be annulled, and the piece of glass which you see, still undisturbed by gravity, to become water. The instantaneous effect of these unequal pressures over its surface will be to set it in motion. If it were a perfect fluid it would go on vibrating for ever with wildly-irregular vibrations, starting from so rude an initial shape as this which I hold in my hand. Water, as any other liquid, is in reality viscous, and therefore the vibrations will gradually

---

[1] This sum for brevity I henceforth call simply "the curvature of the surface" at any point.

C

subside, and the piece of matter will come to rest in a spherical figure, slightly warmed as the result of the work done by the forces of mutual attraction by which it was set in motion from the initial shape. The work done by these forces during the change of the body from any one shape to any other, is in simple proportion to the diminution of the whole surface area; and the configuration of equilibrium, when there is no disturbance from gravity, or from any other solid or liquid body, is that figure—a sphere—in which the surface area is the smallest possible that can enclose the given bulk of matter.

I have calculated the period of vibration of a sphere of water[1] (a dew-drop!) and find it to be $1/4 . a^{3/2}$, where a is the radius measured in centimetres; thus—

For a radius of			the period is	
	1/4 cm.	,,	1/32	second
,,	1 ,,	,,	1/4	,,
,,	2·54 ,,	,,	1	,,
,,	4 ,,	,,	2	,,
,,	16 ,,	,,	16	,,
,,	36 ,,	,,	36	,,
,,	1407 ,,	,,	13,200	,,

[1] See a paper by Lord Rayleigh in *Proc. Roy. Soc.* No. 196, May 5, 1879.

The dynamics of the subject, so far as a single liquid is concerned, is absolutely comprised in the mathematics without symbols which I have put before you. Twenty pages covered with sextuple integrals could tell us no more.

Hitherto we have only considered mutual attraction between the parts of two portions of one and the same liquid — water for instance Consider, now, two different kinds of liquid for instance, water and carbon disulphide (which, for brevity, I shall call sulphide). Deal with them exactly as we dealt with the two pieces of water. I need not go through the whole process again ; the result is obvious. Thirty times the excess of the sum of the surface-tensions of the two liquids separately, above the tension of the interface between them, is equal to the work done in letting the two bodies come together directly over the supposed area of thirty square centimetres. *Hence the interfacial tension per unit area of the interface, is equal to the excess of the sum of the surface-tensions of the two liquids separately, above the*

*work done in letting the two bodies come together
directly so as to meet in a unit area of each.* In
the particular case of two similar bodies coming
together into perfect contact, the interfacial
tension must be zero, and therefore the work
done in letting them come together over a unit
area must be exactly equal to twice the surface-
tension; which is the case we first considered.

If the work done between two different liquids
in letting them come together over a small area,
exceeds the sum of the surface-tensions, the
interfacial tension is negative. The result is an
instantaneous puckering of the interface as the
commencement of diffusion, and the well-known
process of continued inter-diffusion follows.

Consider next the mutual attraction between
a solid and a liquid. Choose any particular
area of the solid, and let a portion of the surface
of the liquid be preliminarily shaped to fit it.
Let now the liquid, kept for the moment rigid,
be allowed to come into contact over this area
with the solid. The amount by which the work
done per unit area of contact falls short of the

surface-tension of the liquid is equal to the interfacial tension of the liquid. If the work done
per unit area is exactly equal to the free-surface
tension of the liquid, the interfacial tension is
zero. In this case the surface of the liquid, when
in equilibrium at the place of meeting of liquid
and solid, is at right angles to the surface of
the solid. The angle between the free surfaces of
liquid and solid is acute or obtuse according as
the interfacial tension is positive or negative;
its cosine being equal to the interfacial tension
divided by the free-surface tension. The greatest
possible value the interfacial tension can have
when positive, is clearly the free-surface tension,
and it reaches this limiting value only in the,
not purely static, case of a liquid resting on a solid
of high thermal conductivity, kept at a temperature
greatly above the boiling-point of the liquid ; as in
the well-known phenomena to which attention has
been called by Leidenfrost and Boutigny. There
is no such limit to the absolute value of the
interfacial tension when negative, but its absolute
value must be less than that of the free-surface

tension to admit of equilibrium at a line of separation between liquid and solid. If minus the interfacial tension is exactly equal to the free-surface tension, the angle between the free surfaces at the line of separation is exactly 180°. If minus the interfacial tension exceeds the free-surface tension, the liquid runs all over the solid, as, for instance, water over a glass plate which has been very perfectly cleansed. If for a moment we leave the centre of the earth, and suppose ourselves anywhere else in or on the earth, we find the liquid running up, against gravity, in a thin film over the upper part of the containing vessel, and leaving the interface at an angle of 180° between the free surface of the liquid, and the surface of the film adhering to the solid above the bounding line of the free liquid surface. This is the case of water contained in a glass vessel, or in contact with a piece of glass of any shape, provided the surface of the glass be very perfectly cleansed.

When two liquids which do not mingle, that is to say, two liquids of which the interfacial tension is positive, are placed in contact and left to themselves undisturbed by gravity (in our favourite laboratory at the centre of the earth suppose), after performing vibrations subsiding in virtue of viscosity, the compound mass will come to rest, in a configuration consisting of two intersecting segments of spherical surfaces constituting the outer boundary of the two portions of liquid, and a third segment of spherical surface through their intersection constituting the interface between the two liquids. These three spherical surfaces meet at the same angles as three balancing forces in a plane, whose. magnitudes are respectively the surface tensions of the outer surfaces of the two liquids and the tension of their interface.   Figs. 2 to 5 (see pages 24, 25) illustrate these configurations in the case of bisulphide of carbon and water for several different proportions of the volumes of the two liquids. [In the figures the dark shading represents water (or sulphate of zinc) in each case.] When the volume of each liquid is given, and the angles of meeting of the three surfaces are

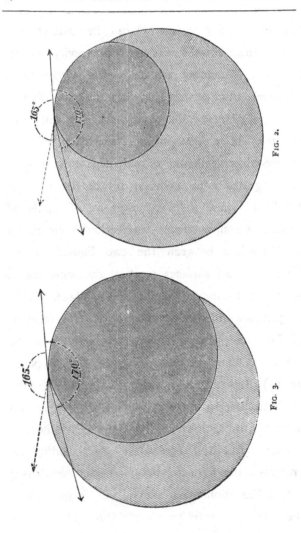

FIG. 2.

FIG. 3.

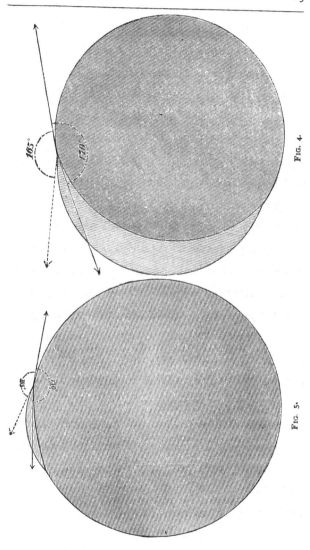

FIG. 4.

FIG. 5.

known, the problem of describing the three spherical surfaces is clearly determinate. It is an interesting enough geometrical problem.

If we now for a moment leave our gravitation-less laboratory, and, returning to the Theatre of the Royal Institution, bring our two masses of liquid into contact, as I now do in this glass bottle, we have the one liquid floating upon the other, and the form assumed by the floating liquid may be learned, for several different cases, from the phenomena exhibited in these bottles and glass beakers, and shown on an enlarged scale in these two diagrams (Figs. 6 to 8, see page 27) ; which represent bisulphide of carbon floating on the surface of sulphate of zinc, and in this case (Fig. 8) the bisulphide of carbon drop is of nearly the maximum size capable of floating. Here is the bottle whose contents are represented in Fig. 8, and we shall find that a very slight vertical disturbance serves to submerge the mass of bisulphide of carbon. There now it has sunk, and we shall find when its vibrations have ceased that the bisulphide of carbon has taken the form of a large

FIG. 6.

FIG. 7.

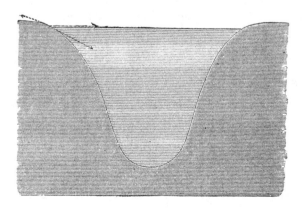

FIG. 8.

sphere supported within the sulphate of zinc. Now, remembering that we are again at the centre of the earth, and that gravity does not hinder us, suppose the glass matter of the bottle suddenly to become liquid sulphate of zinc, this mass would become a compound sphere like the one shown on this diagram (Fig. 3), and would have a radius of about 8 centimetres. If it were sulphate of zinc alone, and of this magnitude, its period of vibration would be about $5\frac{1}{2}$ seconds.

Fig. 9 shows a drop of sulphate of zinc floating on a wine-glassful of bisulphide of carbon.

In observing the phenomena of two liquids in contact, I have found it very convenient to use sulphate of zinc (which I find, by experiment, has the same free-surface tension as water) and bisulphide of carbon ; as these liquids do not mix when brought together, and, for a short time at least, there is no chemical interaction between them. Also, sulphate of zinc may be made to have a density less than, or equal to, or greater than, that of the bisulphide, and the bisulphide may be coloured to a more or less deep purple tint by iodine, and this enables us easily to observe drops of any one of these liquids on the other. In the three bottles now before you the clear

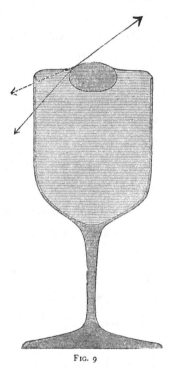

FIG. 9

liquid is sulphate of zinc—in one bottle it has a density less than, in another equal to, and in the third greater than, the density of the sulphide—

and you see how, by means of the coloured sulphide, all the phenomena of drops resting upon or floating within a liquid into which they do not diffuse may be observed, and, under suitable arrangements, quantitatively estimated.

When a liquid under the influence of gravity is supported by a solid, it takes a configuration in which the difference of curvature of the free surface at different levels is equal to the difference of levels divided by the surface tension reckoned in terms of weight of unit bulk of the liquid as unity ; and the free surface of the liquid leaves the free surface of the solid at the angle whose cosine is, as stated above, equal to the interfacial tension divided by the free-surface tension, or at an angle of 180° in any case in which minus the interfacial tension exceeds the free-surface tension. The surface equation of equilibrium and the boundary conditions thus stated in words, suffice fully to determine the configuration when the volume of the liquid and the shape and dimensions of the solid are given. When I say determine, I do not mean unambiguously. There may of course be a multiplicity of solutions of the problem ; as, for instance, when the solid presents several hollows in which, or projections hanging from which, portions of the liquid, or in or hanging from any one of which the whole liquid, may rest.

When the solid is symmetrical round a vertical axis, the figure assumed by the liquid is that of a figure of revolution, and its form is determined by the equation given above in words. A general solution of this problem by the methods of the differential and integral calculus transcends the powers of mathematical analysis, but the following simple graphical method of working out what constitutes mathematically a complete solution, occurred to me a great many years ago.

Draw a line to represent the axis of the surface of revolution. This line is vertical in the realisation now to be given, and it or any line parallel to it will be called vertical in the drawing, and any line perpendicular to it will be called horizontal. The distance between any two horizontal lines in the drawing will be called *difference of levels.*

Through any point, N, of the axis draw a line,

N P, cutting it at any angle (see Fig. 9A). With any point, O, as centre on the line N P, describe a very small circular arc through P P', and let N' be the point in which the line of O P' cuts the

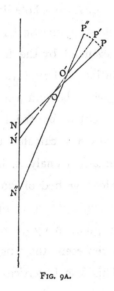

FIG. 9A.

axis. Measure N P, N' P', and the difference of levels between P and P'. Denoting this last by δ, and taking a as a linear parameter, calculate the value of

$$\left(\frac{\delta}{a^2} + \frac{1}{OP} + \frac{1}{NP} - \frac{1}{N'P'}\right)^{-1}.$$

Take this length on the compasses, and putting the pencil point at P', place the other point at O' on the line P' N', and with O' as centre, describe a small arc, P' P''. Continue the process according to the same rule, and the successive very small arcs so drawn will constitute a curved line, which is the generating line of the surface of revolution enclosing the liquid, according to the conditions of the special case treated.

This method of solving the capillary equation for surfaces of revolution remained unused for fifteen or twenty years, until in 1874 I placed it in the hands of Mr. John Perry (now Professor of Mechanics at the City and Guilds Institute), who was then attending the Natural Philosophy Laboratory of Glasgow University. He worked out the problem with great perseverance and ability, and the result of his labours was a series of skilfully executed drawings representing a large variety of cases of the capillary surfaces of revolution. These drawings, which are most instructive and valuable, I have not yet been able to prepare for publication, but the most characteristic of

D

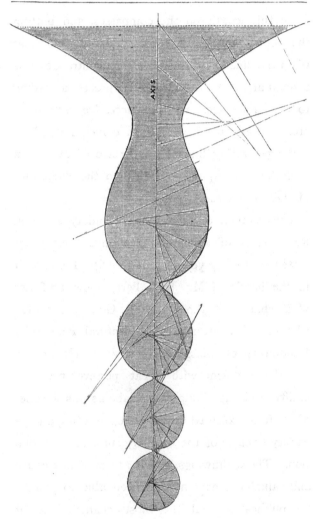

Fig. 10.

them have been reproduced on an enlarged scale, and are now on the screen before you.[1]   Three of

FIG. 11.

the diagrams, those to which I am now pointing (Figs. 10, 11, and 12), illustrate strictly theoretical

[1] The diagrams here referred to were first published in Figs. 10 to 24 of the *Nature* report of this Lecture (July 22 and 29, and August 19, 1886).   These figures are accurate copies of Mr. Perry's original drawings, and I desire to acknowledge the great care and attention which Mr. Cooper, engraver to *Nature*, has given to the work.

solutions—that is to say, the curves there shown do not represent real capillary surfaces—but

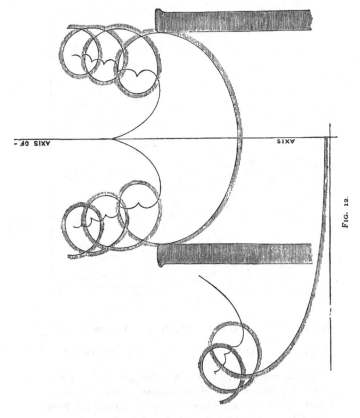

FIG. 12.

such mathematical extensions of the problem, while most interesting and instructive, cannot be adequately treated in the time now at my disposal.

In these other diagrams however, (Figs. 13 to 28,) we have certain portions of the curves taken

FIG. 13.—Mercury in contact with solid spheres (say of glass).

to represent real capillary surfaces shown in section. In Fig. 13 a solid sphere is shown in

FIG. 14.—Sectional view of circular V-groove containing mercury.

four different positions in contact with a mercury surface ; and again, in Fig. 14 we have a section of the form assumed by mercury resting in a circular V-groove. Figs. 15 to 28 (pp. 39-42) show

water-surfaces under different conditions as to
capillarity; the scale of the drawings for each set
of figures is shown by a line the length of which
represents one centimetre; the dotted horizontal
lines indicate the positions of the free water-level.
The drawings are sufficiently explicit to require
no further reference here save the remark that
water is represented by the lighter shading and
solid by the darker.

We have been thinking of our pieces of rigidified
water as becoming suddenly liquified, and conceiv-
ing them enclosed within ideal contractile films;
I have here an arrangement by which I can
exhibit on an enlarged scale a pendant drop,
enclosed not in an *ideal* film, but in a *real* film of
thin sheet india-rubber. The apparatus which
you see here suspended from the roof is a stout
metal ring of 60 centimetres diameter, with its
aperture closed by a sheet of india-rubber tied to
it all round, stretched uniformly in all directions
and as tightly as could be done without special
apparatus for stretching it and binding it to the
ring when stretched.

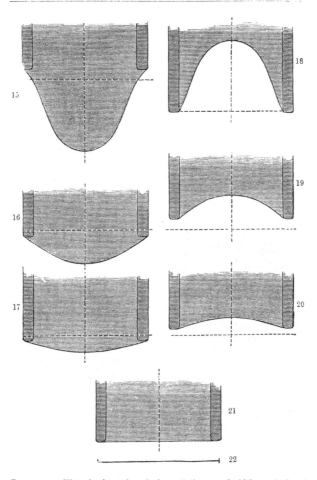

FIGS. 15-21.—Water in glass tubes, the internal diameter of which may be found from Fig. 22, which represents a length of one centimetre. The dotted horizontal line in each figure represents the position of free water-level.

FIG. 23.—Water resting in the space between a solid cylinder and a concentric hollow cylinder.

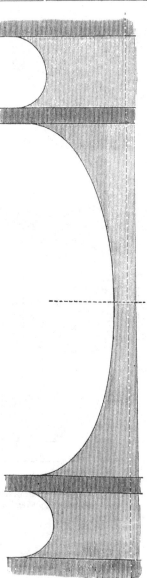

FIG. 24.—Water resting in two co-axial cylinders ; scale is represented by Fig. 28

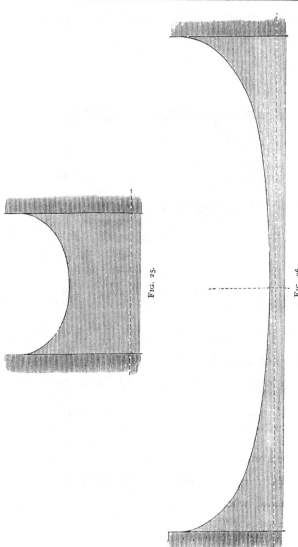

FIG. 25.

FIG. 26.

FIGS. 25 and 26.—Water resting in hollow cylinders (tubes); scale is represented by Fig. 28.

I now pour in water, and we find the flexible bottom assuming very much the same shape as the drop which you saw hanging from my finger after

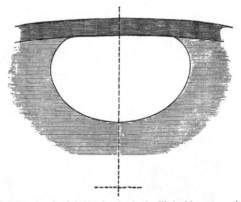

FIG. 27 —Section of the air bubble in a level tube filled with water, and bent so that its axis is part of a circle of large radius ; scale is represented in Fig. 28.

it had been dipped into and removed from the vessel of water (see Fig. 16, p. 39, above). I continue to pour in more water, and the form changes

FIG. 28.—Represents a length of one centimetre for Figs. 24 to 27.

gradually and slowly, preserving meanwhile the general form of a drop such as is shown in Fig. 15, until, when a certain quantity of water has been poured in, a sudden change takes place. The sudden change corresponds to the breaking away of a real drop of water from, for example, the mouth of a tea-urn when the stopcock is so nearly closed that a very slow dropping takes place. The drop in the india-rubber bag, however, does not fall away, because the tension of sheet india-rubber increases enormously when it is stretched. The tension of the real film at the surface of a drop of water remains constant, however much the surface is stretched, and therefore the drop breaks away instantly when enough of water has been supplied from above, to feed the drop to the greatest volume that can hang from the particular size of tube which is used.

I now put this siphon into action, gradually drawing off some of the water, and we find the drop gradually diminishes until a sudden change again occurs and it assumes the form we observed (Fig. 16, p. 39, above) when I first poured in the water. I instantly stop the action of the siphon, and we now find that the great drop has two possible forms of stable equilibrium, with an unstable form intermediate between them. Here is an experimental proof of this statement. With the drop in its higher stable form I cause it to vibrate so as alternately to decrease and increase the axial length, and you see that when the vibrations are such as to cause the increase of length to reach a certain limit there is a sudden change to the lower stable form, and we may now leave the mass performing small vibrations about that lower form. I now increase these small vibrations, and we see that, whenever, in one of the upward (increasing) vibrations, the contraction of axial length reaches the limit already referred to, there is again a sudden change, which I promote by gently lifting with my hands, and the mass assumes the higher stable form, and we have it again performing small vibrations about this form.

The two positions of stable equilibrium, and the one of unstable intermediate between them, is a curious peculiarity of the hydrostatic problem presented by the water supported by india-rubber in the manner of the experiment.

I have here a simple arrangement of apparatus (Figs. 29 and 30) by which, with proper optical aids, such as a cathetometer and a microscope, we can make the necessary measurements on real

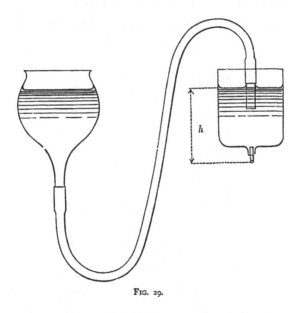

FIG. 29.

drops of water or other liquid, for the purpose of determining the values of the capillary constants. For stability the drop hanging from the open tube should be just less than a hemisphere, but for

convenience it is shown, as in the enlarged drawing of the nozzle (Fig. 30), exactly hemispherical.
By means of the siphon the difference of levels,
h, between the free-level-surface of the water in
the vessel to which the nozzle is attached, and the
lowest point in the drop hanging from the nozzle,

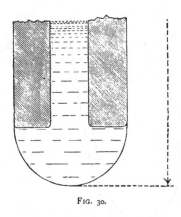

FIG. 30.

may be varied and corresponding measurements
taken of *h*, and of *r*, the radius of curvature of the
drop at its lowest point. This measurement of
the curvature of the drop is easily made with
somewhat close accuracy, by known microscopic
methods The surface tension, T, of the liquid is

CAPILLARY ATTRACTION. 47

calculated from the radius, r, and the observed difference of levels, h, as follows :—

$$\frac{2T}{r} = h \; ;$$

for example, if the liquid taken be water, with a free-surface tension of seventy-five milligrammes per centimetre, and $r = \cdot 05$ cm., h is equal to three centimetres.

Fig. 31, representing a drop of ink just breaking away from the stem of a glass funnel, is drawn from an instantaneous photograph kindly given me by Mr. Graham of Skipness, Argyllshire. He took it himself on an "Ilford quick plate," from a drop of ink just breaking away from the stem of a glass funnel.

Many experiments may be devised to illustrate the effect of surface-tension when two liquids, of which the surface-tensions are widely different, are brought into contact with each other. Thus we may place on the surface of a thin layer of water, wetting uniformly the surface of a glass plate or tray, a drop of alcohol or ether, and so cause the surface-tension of the liquid layer to become smaller in the region covered by the alcohol or ether. On the other hand, from a sur-face-layer of alcohol largely diluted with water, we

FIG. 31.

may arrange to withdraw part of the alcohol at one particular place by promoting its rapid eva-poration, and thereby increase the surface-tension of the liquid layer in that region by diminishing the percentage of alcohol which it contains.

In this shallow tray, the bottom of which is of
ground glass resting on white paper so as to
make the phenomena to be exhibited more easily
visible, there is a thin layer of water coloured
deep blue with aniline; now, when I place on the
water-surface a small quantity of alcohol from this
fine pipette, observe the effect of bringing the
alcohol-surface, with a surface-tension of only
25·5 dynes per lineal centimetre, into contact
with the water-surface, which has a tension of
75 dynes per lineal centimetre. See how the
water pulls back, as it were, all round the alcohol,
forming a circular ridge surrounding a hollow or
small crater, which gradually widens and deepens
until the glass plate is actually laid bare in the
centre, and the liquid is heaped up in a circular
ridge around it. Similarly when I paint with a
brush a streak of alcohol across the tray, we find
the water drawing back on each side from the
portion of the tray touched with the brush.
Now, when I incline the glass tray it is most
interesting to observe how the coloured water
with its slight admixture of alcohol flows down

E

the incline—first in isolated drops, afterwards joining together and forming narrow continuous streams.

These and other well-known phenomena, including that interesting one, "tears of strong wine," were described and explained in a paper "On Certain Curious Motions Observable on the Surfaces of Wine and other Alcoholic Liquors," by my brother, Professor James Thomson, read before Section A of the British Association at the Glasgow meeting of 1855.*

I find that a solution containing about 25 per cent. of alcohol shows the "tears" readily and well, but that they cannot at all be produced if the percentage of alcohol is considerably smaller or considerably greater than 25. In two of those bottles the coloured solution contains respectively one per cent. and 90 per cent. of alcohol; and in them you see it is impossible to produce the "tears"; but when I take this third bottle, in which the coloured liquid contains 25 per cent. of alcohol, and operate upon it, you see—there—the

* See Appendix A to the present Lecture, p. 57 below.

"tears" begin to form at once. I first incline and rotate the bottle so as to wet its inner surface with the liquid, and then, leaving it quite still, I remove the stopper, and withdraw by means of this paper tube the mixture of air and alcoholic vapour from the bottle and allow fresh air to take its place. In this way I promote the evaporation of alcohol from all liquid surfaces within the bottle, and where the liquid is in the form of a thin film it very speedily loses a great part of its alcohol. Hence the surface tension of the thin film of liquid on the interior wall of the bottle comes to have a greater and greater value than the surface-tension of the mass of liquid in the bottom, and where these two liquid surfaces having different surface-tensions come together, we have the phenomena of "tears." There—as I hasten the evaporation, you see a horizontal ring of liquid being formed and creeping up the side of the bottle: afterwards we find the liquid so raised collecting into drops which slip down the side and give a fringe-like appearance to

E 2

the space through which the rising ring has passed.*

These phenomena may also be observed by using, instead of alcohol, ether, which has a sur-face-tension equal to about three-fourths of that of alcohol. In using ether, however, this very curious effect may be seen.† I dip the brush into the ether, and hold it near to but not touching the water-surface. Now I see a hollow formed, which becomes more or less deep according as the brush is nearer to or farther from the normal water-surface, and the hollow follows the brush about as I move it so.

Here is an experiment showing the effect of heat

---

* The following paragraph, quoted from Clerk-Maxwell's "Heat," Ed. 1871, p. 273, contains an interesting reference to this part of our subject.

"This phenomenon, known as the tears of strong wine, was first explained on these principles by Professor James Thomson. It is probable that it is referred to in Proverbs xxiii. 31, as an indication of the strength of the wine. The motion ceases in a stoppered bottle as soon as enough of vapour of alcohol has been formed in the bottle to be in equilibrium with the liquid alcohol in the wine."

† See Clerk-Maxwell's article (p. 65) on "Capillary Attraction" ("Encyclopædia Britannica," 9th edition).

on surface-tension. Over a portion of this tin plate there is a thin layer of resin. I lay the tin plate on the top of this hot copper cylinder, when we at once see the fluid resin drawing back from the portion of the plate directly over the end of the heated copper cylinder, and leaving a circular space on the surface almost clear of resin, showing how very much the surface-tension of hot resin is less than that of cold resin.

Note of January 30, 1886.—The equations (8) and (9) on p. 59 of Clerk-Maxwell's article, on " Capillary Attraction " in the ninth edition of the " Encyclopædia Britannica " do not contain terms depending on the mutual action between the two liquids, and the concluding expression (10), and the last small print paragraph of the page are wholly vitiated by this omission. The paragraph immediately following equation (10) is as follows :—

" If this quantity is positive, the surface of contact will tend to contract, and the liquids will remain distinct. If, however, it were negative, the displacement of the liquids which tends to enlarge the surface of contact would be aided by the molecular forces, so that the liquids, if not kept separate by gravity, would become thoroughly mixed. No instance, however, of a phenomenon of this kind has been discovered, for those liquids which mix of themselves do so by the process of diffusion, which is a molecular motion, and not by the spontaneous puckering and replication of the boundary surface as would be the case if T were negative."

It seems to me that this view is not correct; but that on the contrary there is this "puckering" as the *very beginning* of diffusion. What I have given in the lecture as reported in the text above seems to me the right view of the case as regards diffusion in relation to interfacial tension.

It may also be remarked that Clerk-Maxwell, in the large print paragraph of p. 59, preceding equation (I), and in his application of the term potential energy to E in the small print, designated by *energy* what is in reality exhaustion of energy or negative energy; and the same inadvertence renders the small print paragraph on p. 60 very obscure. The curious and interesting statement at the top of the second column of p. 63, regarding a drop of carbon disulphide in contact with a drop of water in a capillary tube would constitute a perpetual motion if it were true for a tube not first wetted with water through part of its bore—" . . . if a drop of water and a drop of bisulphide of carbon be placed in contact in a horizontal capillary tube, the bisulphide of carbon will chase the water along the tube."

Additional Note of June 5, 1886.—I have carefully tried the experiment referred to in the preceding sentence, and have not found the alleged motion.

APPENDIX A.

ON CERTAIN CURIOUS MOTIONS OBSERVABLE ON
THE SURFACES OF WINE AND OTHER ALCO-
HOLIC LIQUORS.

[*A paper by Professor James Thomson, read before Section A
of the British Association at the Glasgow Meeting of
1855: Brit. Assoc. Report for 1855, Part II. pp. 16, 17.*]

THE phenomena of capillary attraction in
liquids are accounted for, according to the
generally received theory of Dr. Young, by the
existence of forces equivalent to a tension of the
surface of the liquid, uniform in all directions,
and independent of the form of the surface. The
tensile force is not the same in different liquids.
Thus it is found to be much less in alcohol than in
water. This fact affords an explanation of several
very curious motions observable, under various
circumstances, at the surfaces of alcoholic liquors.
One part of these phenomena is, that if, in the
middle of the surface of a glass of water, a small
quantity of alcohol, or strong spirituous liquor, be

gently introduced, a rapid rushing of the surface is found to occur outwards from the place where the spirit is introduced. It is made more apparent if fine powder be dusted on the surface of the water. Another part of the phenomena is, that if the sides of the vessel be wet with water above the general level surface of the water, and if the spirit be introduced in sufficient quantity in the middle of the vessel, or if it be introduced near the side, the fluid is even seen to ascend the inside of the glass until it accumulates in some places to such an extent that its weight preponderates and it falls down again. The manner in which the author explains these two parts of the phenomena is, that the more watery portions of the entire surface, having more tension than those which are more alcoholic, drag the latter briskly away, sometimes even so as to form a horizontal ring of liquid high up round the interior of the vessel, and thicker than that by which the interior of the vessel was wet. Then the tendency is for the various parts of this ring or line to run together to those parts which happen to be most watery, and so there is no stable equilibrium ; for, the parts to which the various portions of the liquid aggregate themselves soon become too heavy to be sustained, and so they fall down. The same mode of explanation, when carried a step further, shows the reason of the curious motions commonly observed in the film of wine adhering to the inside of a wine-glass, when the glass, having been partially filled with wine, has been shaken so as to wet the inside above the general level of the surface of the liquid ; for, to explain these motions, it is only necessary further to bring under consideration, that the thin film adhering to the inside of the glass must very quickly become more watery than the rest on account of the evaporation of the alcohol contained in it being more rapid than the evaporation of the water. On this matter, the author exhibited to the Section a very decisive experiment. He showed that in a vial partly filled with wine, no motion of the kind described occurs as long as the vial is kept corked. On his removing the cork, however, and withdrawing by a tube the air saturated with vapour of the wine, so that it was replaced by fresh air capable of producing evaporation, a liquid film was instantly seen as a horizontal ring creeping up the interior of the vial, with viscid-looking pendent streams descending from it like a fringe from a curtain. He gave another striking illustration by pouring water on a flat silver tray, previously carefully cleaned from any film which could hinder the water from thoroughly wetting the surface. The water was about one-tenth of an inch deep. Then, on a little alcohol being laid down in the middle of the tray, the water immediately rushed away from the middle, leaving a deep hollow there, which laid the tray bare of all liquid, except an exceedingly

thin film. These and other experiments, which he
made with fine lycopodium powder dusted on the
surface of the water, into the middle of which he
introduced alcohol gently from a fine tube, were
very simple, and can easily be repeated. Certain
curious return currents which he showed by means
of the powder on the surface, he stated he had not
yet been able fully to explain. He referred to
very interesting phenomena previously observed
by Mr. Varley, and described in the fiftieth volume
of the Transactions of the Society of Arts, and he
believed that many or all of these would prove to
be explicable according to the principles he had
now proposed.

APPENDIX B.

NOTE ON GRAVITY AND COHESION.

[*A paper read before the Royal Society of Edinburgh and
published in Proc. R. S. E. April 21, 1862.*]

THE view, founded on Boscovich's theory, commonly taken of cohesion, whether of solids or
liquids, is, that it results from a force of attraction between the particles of matter, which
increases much more rapidly than according to

the inverse square of the distance, when the distance is diminished below some very small limit. This view might, indeed, seem inevitable, unless the idea of "attraction" is to be discarded altogether; because the law of attraction at sensible distances—the Newtonian law—demonstrated by its discoverer for distances not incomparably smaller than the earth's dimensions, and verified by Maskelyne and Cavendish in a manner rendering it impossible for any naturalist to reasonably doubt its applicability to the mutual action between particles a few hundred yards or a few inches asunder, seems to give only very small, scarcely appreciable, forces between bodies of such masses as those we experiment on in our laboratories, everywhere placed as close as possible to one another—that is to say, in contact; and does not seem to provide for any considerable increase of attraction when the area of contact is increased, whether by pressing the bodies together, or by shaping them to fit over a large area.

But if we take into account the heterogeneous distribution of density essential to any molecular theory of matter, we readily see that it alone is sufficient to intensify the force of gravitation between two bodies placed extremely close to one another, or between two parts of one body, and therefore that cohesion may be accounted for without assuming any other force than that of gravitation, or any other law than the Newtonian. To prove this, let two homogeneous cubes be placed with one side of each in perfect contact with one side of the other; and let one-third of the matter of each cube be condensed into a very great number, i, of square bars perpendicular to the common face of the two; and let the other two-thirds of the matter be removed for the present. The mass of each bar will be $1/3i$ of the whole mass originally given in each cube.

Let us farther suppose that the two groups of bars are placed so that each bar of one group has an end in complete contact with an end of a bar of the other. The attraction between each two such conterminous bars, however small their masses are, may be increased without limit, by diminishing the area of its section, and keeping its mass constant. But the whole mutual attraction between the two groups exceeds i times the attraction between each of the conterminous pairs, and may therefore be made to have any value, however great, merely by condensing each bar in its transverse section, and keeping their number and the mass of each constant.

We may now suppose another third of the whole mass to be condensed into bars parallel to another side of the cube, and the remaining third into bars parallel to the remaining side. If, then, either of these cubes be placed with any side in

contact with any side of the other, and allowed
to take the relative position to which it will
obviously tend—that in which the bars perpendicular to the common sides of the two cubes
come together end to end—there will be produced,
by pure gravitation, a force of attraction between
them which may be of any amount, however
great, and which will be the greater the greater
the ratio of the whole space unoccupied within the
boundary of either cube, to the space occupied
by the matter of the bars.

The illustration has been chosen merely for the
sake of definiteness and simplicity; but it is clear
that any arrangement, however complex, of woven
fibrous structure, provided only the ratio of the
unoccupied to the occupied space is sufficiently
great, will lead to the same general conclusion.
Farther, it is clear that the same result would be
produced by any sufficiently intense heterogeneousness of structure whatever, provided only some
appreciable proportion of the whole mass is so
condensed in a continuous space in the interior
that it is possible, from any point of this space
as centre, to describe a spherical surface which
shall contain a very much greater amount of
matter than the proportion of the whole matter
of the body which would correspond to its volume.
Except in imposing this condition, the theory
now suggested interferes with no molecular hypothesis hitherto propounded, continuous or atomic,

finite atoms, or centres of force, static or kinetic.

Physical science abounds with evidence that there is an ultimate very intense heterogeneousness in the constitution of matter. All that is valid of the unfortunately so-called "atomic" theory of chemistry seems to be an assumption of such heterogeneousness in explaining the combination of substances. This alone, it is true, does not explain the law of definite combining proportions; but neither does the hypothesis of infinitely strong finite pieces of matter; and whatever is assumed to be the structural character of a chemical compound, a dynamical law of affinity between the two substances, according to the proportions of them lying or moving beside one another, must be added to do what some writers seem to suppose done by their "atomic theory."

It is satisfactory to find that, so far as cohesion is concerned, no other force than that of gravitation need be assumed.

APPENDIX C.

ON THE EQUILIBRIUM OF VAPOUR AT A CURVED SURFACE OF LIQUID.

[*A paper read before the Royal Society of Edinburgh and published in Proc. R. S. E. February 7th,* 1870 (vol. vii. pp. 63-68).*]

IN a closed vessel containing only a liquid and its vapour, all at one temperature, the liquid rests, with its free surface raised or depressed in capillary tubes and in the neighbourhood of the solid boundary, in permanent equilibrium according to the same law of relation between curvature and pressure as in vessels open to the air. The permanence of this equilibrium implies physical equilibrium between the liquid and the vapour in contact with it at all parts of its surface. But the pressure of the vapour at different levels differs according to hydrostatic law. Hence the pressure of saturated vapour in contact with a liquid differs according to the curvature of the bounding surface, being less when the liquid is concave, and greater when it is convex. And detached portions of the liquid in separate vessels all enclosed in one containing vessel, cannot remain permanently with their free surfaces in any other relative positions

than those they would occupy if there were hydrostatic communication of pressure between the
portions of liquid in the several vessels. There
must be evaporation from those surfaces which are
too high, and condensation into the liquid at those
surfaces which are too low—a process which goes
on until hydrostatic equilibrium, as if with free
communication of pressure from vessel to vessel,
is attained. Thus, for example, if there are two
large open vessels of water, one considerably above
the other in level, and if the temperature of the
surrounding matter is kept rigorously constant, the
liquid in the higher vessel will gradually evaporate
until it is all gone and condensed into the lower
vessel. Or if, as illustrated by the annexed diagram
(Fig. 32), a capillary tube, with a small quantity of
liquid occupying it from its bottom up to a certain
level, be placed in the neighbourhood of a quantity
of the same liquid with a wide free surface, vapour
will gradually become condensed into the liquid in
the capillary tube until the level of the liquid in it
is the same as it would be were the lower end of
the tube in hydrostatic communication with the
large mass of liquid. Whether air be present
above the free surface of the liquid in the several
vessels or not, the condition of ultimate equilibrium
is the same; but the processes of evaporation
and condensation through which equilibrium is
approached will be very much retarded by the
presence of air. The experiments of Graham,

F

and the kinetic theory of Clausius and Maxwell, scarcely yet afford us sufficient data for estimating the rapidity with which the vapour proceeding from one of the liquids will diffuse itself through the air and reach the surface of another liquid at a

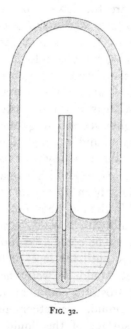

Fig. 32.

lower level. With air at anything approaching to ordinary atmospheric density to resist the process, it is probable it would be too slow to show any results unless in very long continued experiments. But if the air be removed as perfectly as can be done by well-known practical methods, it is probable that the process will be very rapid : it would, indeed, be instantaneous, were it not for the cold of evaporation in one vessel and the heat of condensation in the other. Practically, then, the rapidity of the process towards hydrostatic equilibrium through vapour, between detached liquids, depends on the rate of the conduction of heat between the several surfaces through intervening solids and liquids. Without having made either the experiment, or any calculations on the rate of conduction of heat in the circumstances, I feel convinced that in a very short time water would visibly rise in the capillary tube indicated in the diagram (Fig. 32), and that, provided care is taken to maintain equality of temperature all over the surface of the hermetically-sealed vessel, the liquid in the capillary tube would soon take very nearly the same level as it would have were its lower end open ; sinking to this level if the capillary tube were in the beginning filled too full, or rising to it if (as indicated in the diagram) there is not enough of liquid in it at first to fulfil the condition of equilibrium.

The following formulæ show precisely the relations between curvatures, differences of level, and differences of pressure, with which we are concerned.

Let ρ be the density of the liquid, and σ that of the vapour ; and let T be the cohesive tension of the free surface, per unit of breadth, in terms of weight of unit mass, as unit of force. Let h denote the height of any point, P, of the free surface above a certain plane of reference, which I shall call for brevity the plane level of the free surface. This will be sensibly the actual level of the free surface in regions, if there are any, with no part of the edge (or bounding line of the free surface where liquid ends and solid begins) at a less distance than several centimetres. Lastly, let r and r' be the principal radii of curvature of the surface at P. By Laplace's well-known law, we have, as the equation of equilibrium,

$$(\rho - \sigma)h = \mathrm{T}\left(\frac{1}{r} + \frac{1}{r'}\right). \qquad (1).$$

Now, in the space occupied by vapour, the pressure is less at the higher than at the lower of two points whose difference of levels is h, by a difference equal to σh. And there is permanent equilibrium between vapour and liquid at all points of the free surface. Hence the pressure of vapour in equilibrium is less at a concave than at a plane surface of liquid, and less at a plane surface than at a convex surface, by differences amounting to $\dfrac{\mathrm{T}\sigma}{\rho - \sigma}$ per unit difference of curvature. That is to say, if ϖ denote the pressure of vapour in equilibrium at a plane surface of liquid, and p the pressure of vapour of the same liquid at the same temperature presenting a curved surface to the vapour, we have

$$p = \varpi - \frac{\mathrm{T}\sigma}{\rho - \sigma}\left(\frac{1}{r} + \frac{1}{r'}\right). \qquad (2).$$

$\frac{1}{r}$ and $\frac{1}{r'}$ being the curvatures in the principal sections of the surface bounding liquid and vapour,
reckoned positive when concave towards the
vapour.

In strictness, the value of σ to be used in these
equations, (1) and (2), ought to be the mean
density of a vertical column of vapour, extending
through the height h from the plane of reference.
But in all cases to which we can practically apply
the formulæ, according to present knowledge of
the properties of matter, the difference of densities
in this column is very small, and may be neglected.
Hence, if H denote the height of an imaginary
homogeneous fluid above the plane of reference,
which, if of the same density as the vapour at that
plane, would produce by its weight the actual
pressure ϖ, we have

$$\sigma = \frac{\varpi}{H}.$$

Hence by (1) and (2)

$$p = \varpi \left(1 - \frac{h}{H} \right) \qquad . \qquad . \qquad (3).$$

For vapour of water at ordinary atmospheric
temperatures, H is about 1,300,000 centimetres.
Hence, in a capillary tube which would keep
water up to a height of 13 metres above the plane
level, the curved surface of the water is in equilibrium with the vapour in contact with it, when
the pressure of the vapour is less by about $\frac{1}{1000}$th

of its own amount than the pressure of vapour in equilibrium at a plane surface of water at the same temperature.

For water the value of T at ordinary temperatures is about o8 of a gramme weight per centimetre ; and ρ, being the mass of a cubic centimetre, in grammes, is unity. The value of σ for vapour of water at any atmospheric temperature is so small that we may neglect it altogether in equation (1). In a capillary tube thoroughly wet with water, the free surface is sensibly hemispherical, and therefore r and r' are each equal to the radius of the inner surface of the liquid film lining the tube above the free liquid surface ; we have, therefore,

$$h = \text{·08} \times \frac{2}{r}.$$

Hence, if $h = 1300$ centimetres, $r = \text{·00012}$ centimetres. There can be no doubt but that Laplace's theory is applicable without serious modification even to a case in which the curvature is so great (or radius of curvature so small) as this. But in the present state of our knowledge we are not entitled to push it much further. The molecular forces assumed in Laplace's theory to be "insensible at sensible distances" are certainly but little, if at all, sensible at distances equal to or exceeding the wave lengths of ordinary light. This is directly proved by the most cursory observation of soap bubbles. But the appearances

presented by the black spot which abruptly ends
the series of colours at places where the bubble is
thinnest before it breaks, make it quite certain that
the action of those forces becomes sensible at
distances not much less than a half wave length,
or 1/40000 of a centimetre. There is, indeed, much
and multifarious evidence that in ordinary solids
and liquids, not merely the distances of sensible
inter-molecular action, but the linear dimensions
of the molecules themselves, and the average
distance from centre to nearest centre,[1] are but
very moderately small in comparison with the
wave lengths of light. Some approach to a definite estimate of the dimensions of molecules is
deducible from Clausius' theory of the average
spaces travelled without collision by molecules of
gases, and Maxwell's theory and experiments regarding the viscosity of gases. Having perfect
confidence in the substantial reality of the views
which these grand investigations have opened to
us, I find it scarcely possible to admit that there
can be as many as 10^{27} molecules in a cubic centimetre of liquid carbonic acid or of water. This
makes the average distance from centre to nearest
centre in the liquids exceed a thousand-millionth
of a centimetre!

We cannot, then, admit that the formulæ which

---

[1] By "average distance from centre to nearest centre," I mean
the side of the cube in a cubic arrangement of a number of points
equal to the number of real molecules in any space.

I have given above are applicable to express the
law of equilibrium between the moisture retained
by vegetable substances, such as cotton cloth or
oatmeal, or wheat-flour biscuits, at temperatures
far above the dew point of the surrounding
atmosphere. But although the energy of the
attraction of some of these substances for vapour
of water (when, for example, oatmeal, previously
dried at a high temperature, has been used, as in
the original experiment of Sir J. Leslie, to produce
the freezing of water under the receiver of an air-
pump) is so great that it might almost claim re-
cognition from chemists as due to a " chemical
affinity," and resulting in a " chemical combination,"
I believe that the absorption of vapour into fibrous
and cellular organic structures is a property of
matter continuous with the absorption of vapour
in a capillary tube demonstrated above.

ELECTRICAL UNITS OF MEASUREMENT.

[*A Lecture delivered at the Institution of Civil Engineers on May* 3, 1883 ; *being one of a series of Six Lectures on "The Practical Applications of Electricity."*]

IN physical science a first essential step in the direction of learning any subject is to find principles of numerical reckoning and methods for practicably measuring some quality connected with it. I often say that when you can measure what you are speaking about, and express it in numbers, you know something about it ; but when you cannot measure it, when you cannot express it in numbers, your knowledge is of a meagre and unsatisfactory kind : it may be the beginning of knowledge, but you have scarcely, in your thoughts, advanced to the stage of *science*, whatever the matter may be. I may illustrate by a case in which this first step has not been taken. The hardness of different solids, as precious stones and metals, is reckoned by a merely comparative test. Diamond cuts ruby, ruby cuts quartz, quartz, I believe, cuts glass-hard steel, and glass-hard steel cuts glass ; hence diamond is reckoned harder than ruby ; ruby, than quartz ; quartz, than glass-hard steel ; and glass-hard steel, than glass : but we have no numerical measure of the hardness of these, or of any other solids. We have, indeed, no knowledge of the moduluses of rigidity, or of the tensile strength, of almost any of the gems or minerals, of which the hardness is reckoned by mineralogists in their comparative scale, beginning with diamond, the hardest of known solids. We have even no reason to believe that the modulus of rigidity of diamond is greater than that of other solids ; and we have no exact understanding of what this property of hardness is, nor of how it is related to moduluses of elasticity, or to tensile or shearing strength, or to the quality of the substance in respect to its bearing stresses exceeding the limit of its elasticity. It must, therefore, be admitted, that the science of strength of materials, so all-important in engineering, is but little advanced, and the part of it relating to the so-called hardness of different solids least of all; there being in it no step toward quantitative measurement or reckoning in terms of a definite unit.

A similar confession might have been made regarding electric science, as studied even in the chief physical laboratories of the world, ten years ago. True, Cavendish and Coulomb last century, and Ampère, and Poisson, and Green, and Gauss, and Weber, and Ohm, and Lentz, and Faraday, and Joule, this century, had given us the mathematical and experimental foundation for a complete system of numerical reckoning in electricity and magnetism, in electro-chemistry, and in electro-thermodynamics ; and as early as 1858 a practical beginning of definite electric measurement had been made, in the testing of copper resistances, insulation resistances, and electro-static inductive capacities of submarine cables. But fifteen years passed after this beginning was made, and resistance

coils and ohms, and standard condensers and micro-
farads, had been for ten years familiar to the
electricians of the submarine-cable factories and
testing-stations, before anything that could be called
electric measurement had come to be regularly
practised in most of the scientific laboratories of
the world. I doubt whether, ten years ago, a single
scientific-instrument maker or seller could have
told his customers whether the specific conductivity
of his galvanometer coils was anything within 60
per cent. of that of pure copper ; and I doubt
whether the resistances of one in a hundred of the
coils of electro-magnets, galvanometers, and other
electro-magnetic apparatus, in the universities, and
laboratories, and lecture establishments of the world,
were known to the learned professors whose duty it
was to explain their properties, and to teach their
use to students and pupils. But we have changed
all that ; and now we know the resistances of our
electro-magnetic coils, generally speaking, better
than we know their lengths ; and our least advanced students in physical laboratories are quite
able to measure resistances through a somewhat

wide range with considerable accuracy. I should
think, indeed, that with the appliances in ordinary
use, they are more likely to measure resistances of
from 100 to 10,000 ohms to an accuracy of $\frac{1}{1}$ per
cent., than they are to be right to one millimetre in a
metre in their measurements of length. It certainly
is a very surprising result that in such a recondite
phenomenon—such a subtle quality to deal with—
as electric resistance, which is so very difficult to
define, and which we are going to learn is a velocity,
every clerk in a telegraph station, the junior
students and assistants in laboratories, and even
workmen in electric lighting establishments, are
perfectly ready to measure, more accurately than
you would measure the length of ten feet of wire,
the resistance of electric conductors in definite
absolute units.

I suppose, too, nearly every apparatus-room and
physical laboratory possesses a micro-farad, but 1
am afraid its pedigree is not often known ; and if
its accuracy within 10 per cent. were challenged, I
doubt whether, in many cases, any one, whether
maker, or possessor, or other electrical expert,

could be found to defend it. As for our electrostatic apparatus, I confess that I do not know the
capacity of a single one of the two or three dozen
Leyden jars, which in 1846 I inherited, in the
Natural Philosophy apparatus-room of the
University of Glasgow, or which I have made
from time to time during the thirty-seven years
passed since that date. I would fain hope that I
am singular in such a confession, and that no
other professor of Natural Philosophy in the
world would let a Leyden jar be put on his
lecture-room table without being able to tell
his students its capacity in absolute measure.
The reckoning of Leyden-jar capacity in square
inches of coated glass—thickness and specific
inductive capacity not stated—ought to be as
much a thing of the past as is the reckoning of
resistances in terms of a mile of wire—weighing
fourteen grains to the foot, of ordinary
commercial copper, specific resistance not stated
—perhaps 45 per cent.? or 70 per cent.? or 98
per cent.? of the conductivity of pure copper.
And as to practical measurement of electromotive

force, we have scarcely emerged one year from those middle ages when a volt and a Daniell's cell were considered practically identical, to the higher aspiration of measurement within one per cent. It seems, indeed, as if the commercial requirements of the application of electricity to lighting, and other uses of every-day life, were destined to cause an advance of the practical science of electric measurement, not less important and valuable in the higher region of scientific investigation than that which, from twenty to thirty years ago, was brought about by the practical requirements of submarine telegraphy.

There cannot be a greater mistake than that of looking superciliously upon practical applications of science. The life and soul of science is its practical application, and just as the great advances in mathematics have been made through the desire of discovering the solution of problems which were of a highly practical kind in mathematical science, so in physical science many of the greatest advances that have been made from the beginning of the world to the present time have been made in the earnest desire to turn the knowledge of the properties of matter to some purpose useful to mankind.

The first step toward numerical reckoning of properties of matter, more advanced than the mere reference to a set of numbered standards, as in the mineralogist's scale of hardness, or to an arbitrary trade standard, as in the Birmingham wire-gauge, is the discovery of a continuously-varying action of some kind, and the means of observing it definitely, and measuring it in terms of some arbitrary unit or scale division. But more is necessary to complete the science of measurement in any department; and that is the fixing on something absolutely definite as the unit of reckoning, which, with reference to electric and magnetic science, is the subject of my lecture of this evening.

In electricity, the mathematical theory and the measurements of Cavendish, and in magnetism, the measurements of Coulomb, gave, one hundred years ago, the requisite foundation for a complete system of measurement: and fifty years ago the same thing was done for electro-magnetism by Ampère.

I speak of electricity, of magnetism, and of electro-magnetism. Now I must premise, as a matter of importance in respect of some of the technical details which we shall have to consider later, that magnetism must be held to include electro-magnetism. Electro-magnetism and magnetism are one and the same thing. Electro-magnetic and electro-static force, which are very distinct just now, are two things which deeper science may lead us to unite, in a manner that we can scarcely see at present. We have the foundation of Cavendish for electricity, of Coulomb for magnetism, and of Ampère for electro-magnetism, which fall in perfectly with what I shall have to say later on, in respect of Gauss and Weber's work, of magnetism and electro-magnetism. I say this, because there has been some little discussion in respect to the magnetic unit and the electro-magnetic unit, as if the magnetic unit might be something different from the electro-magnetic unit, or the electro-

G

kinetic unit. It will simplify matters if we think merely of a magnetic force, whether it be due to a steel magnet, or to a wire conveying a current ; and make no distinction so far as measurement is concerned, through the range of the science of magnetism, including electro-magnetism. We shall find that we have the two capital subjects ; electricity and electro-static force one of them : magnetism and electricity in motion through conductors, and magnetic and electro-magnetic force, the other. The first complete method of scientific measurement for any of these subjects was that of Gauss, in his system of absolute measurement for terrestrial magnetism so splendidly realised by Gauss and Weber in their Magnetic Society of Göttingen, which gave the starting impulse for the whole system of absolute measurement as we now have it, throughout the range of electric science. In fact, Weber himself, after realising absolute measure in terrestrial magnetism in conjunction with Gauss, carried it on through the field of electro-magnetism in his *Elektrodynamische*

Maasbestimmungen,[1] and thence into electrostatics in his joint work with Kohlrausch, under the same title, *Elektrodynamische Maasbestimmungen*.[2] The now celebrated "*v*" (velocity), which Maxwell in his electro-magnetic theory of light pointed out to be not merely by chance approximately equal to the velocity of light, but to be probably connected physically, in virtue of the forces concerned, with the actual action or motion of matter which constitutes light, was found to be approximately 300,000 kilometres per second.[3]

As early as 1851 I commenced using the absolute system in the reckoning of electromotive forces of voltaic cells, and the electric resistances

---

[1] Leipsig, 1852. An earlier publication of one of the most important parts of the work was Weber's paper, "Messungen galvanischen Leitungswiderstände nach einen absoluten Maasse." Poggendorff's *Annalen*, March 1851.

[2] Poggendorff's *Annalen*, August 10, 1856.

[3] The exact number given by Weber and Kohlrausch is 310,740 ; but more recent investigations render it probable that this number may be 3 or 4 per cent. too great. See also Gray's *Absolute Measurements in Electricity and Magnetism* (Macmillan and Co., London, 1883).

of conductors, in absolute electro-magnetic units ; [1] and after advocating the general use of the absolute system, both for scientific investigation and for telegraph work, for ten years, I obtained in 1861 the appointment of a Committee [2] of the British Association on Electrical Standards.

This committee worked for nearly another ten years through the whole field of electro-magnetic and electro-static measurement, but chiefly on standards of electric resistance, until in its final report, presented to the Exeter meeting in August 1869, it fairly launched the absolute system for general use; with arrangements for the supply of standards for resistance coils in terms of a

---

[1] See my papers "On the Mechanical Theory of Electrolysis," and "Applications of the Principle of Mechanical Effect to the Measurement of Electromotive Forces, and of Galvanic Resistances in Absolute Units," both published in the *Philosophical Magazine*, December 1851 ; now constituting Articles LIII. and LIV. of my Reprint of *Mathematical and Physical Papers*," Vol. I., 1882.

[2] The Reports of this Committee were published at intervals from 1861 to 1869 in the British Association volumes of Reports for the respective years.   These, along with other contributions to the subject, were collected and, under the editorship of Professor Fleeming Jenkin, published by Spon, London and New York, 1871.

unit, first called the British Association unit, and afterwards the ohm; of which the resistance reckoned in electro-magnetic measure was to be, as nearly as possible, 10,000 kilometres per second.

In regard to the name of "ohm," I may mention that a paper was communicated to the British Association in 1861 by Sir Charles Bright and Mr. Latimer Clark, in which the names that we now have, with some slight differences, were suggested ; and a complete continuous system of measurement was proposed, which did not fulfil certainly all the conditions of the absolute system, but which fulfilled some of them in an exceedingly useful manner for practical purposes. To Sir Charles Bright and Mr. Latimer Clark, therefore, is due the whole system of names as we have it now, ohms, volts, farads, and micro-farads. From 1870 or 1871 forward, the absolute system, with the approach to accurate realisation of it given by the British Association unit, has been in general use in England and America ; but another decade has passed, a rather long one, before the definitive practical adoption of the absolute system by France, Germany, and other European countries, as decreed by the International Conference for the determination of electric units, held at Paris in October 1882. The decision adopted was, not to take the British Association unit. Doubt was thrown upon its accuracy, which we shall see was well founded. The question of a strict foundation for a metrical system was before the Conference, and it was inclined to adopt the absolute system, but the question occurred " What is the ohm ? " Who can see an ohm ? Who can show what an ohm is ? Who can measure the resistance of any conductor for us, in this absolute measure of Weber's ? Weber's own measurement differed greatly from that of the British Association. Several experimenters, in endeavouring to verify or test the British Association measurement arrived at results which were discordant among themselves, and therefore could not be confirmatory of the British Association measurement. Things were in this doubtful state, and the Conference had a very im-

portant practical question to decide. A proposal
had been before the world for ten years at least, to
found accurate measurement of electrical resistance
upon a material obtainable in uniform quality
and by easy precautions in a state of perfect
purity, or sufficiently nearly perfect to fit it
practically for the purpose in question, which
is,—the giving of a standard for the measurement
of resistance. The Siemens unit, founded upon
the specific resistance of mercury, had been proposed. The great house of Siemens (Berlin and
London), our distinguished *confrère*, Sir William
Siemens, and his distinguished brother, Dr. Werner
Siemens, worked upon this subject in the most
thorough and powerful way—the measurement of
resistances in terms of the specific resistance of
mercury—in such a manner as to give us a standard
which shall be reproducible at any time and place,
·with no other instrument of measurement at hand
than the metre measure. I say, the system of
measurement of resistance on a mercury standard
had been worked out, and its practicability demonstrated. Werner and William Siemens themselves

were both present at the Conference, and they joined heartily in the proposal to adopt an absolute system, but the question was how to make a beginning; and the answer adopted by the Conference was to ask for a definition of an absolute system in terms of a column of mercury. The column of mercury was the one standard in existence, that could be reproduced otherwise than by merely copying from one wire to another; and it was naturally adopted as the foundation upon which a standard, if not a practical unit to be used, should be founded. In short, then, the finding of the Conference was to this effect: that as soon as good evidence is given of a sufficiently near measurement for practical purposes, of the resistance of any con- ductor—be it a piece of wire or a column of mercury—as soon as such measurement should be made, with evidence that it is accurate enough for practical purposes, then the unit which the British Association had aimed at should be adopted; but it was to be left to the judgment and the convenience of the users of standards when to make the change, should a change be necessary from the British Association unit as the ohm, or from the Siemens unit, to bring measurement into more close agreement with the absolute reckoning. What had been done by Lord Rayleigh and Mrs. Sidgwick had left very little room for doubt but that the British Association unit was in error to the extent of 1·3 per cent. The Siemens unit had the advantage of being somewhat approximately equal to the desired absolute unit, though not professing to be an absolute unit at all. It was simply the resistance of a column of mercury at zero temperature, a metre in length and a square millimetre in section. There were great difficulties in the reproduction of the Siemens unit, in the earlier times of the investigation; but Dr. Werner Siemens, and Lord Rayleigh, and Mrs. Sidgwick, and many other workers besides, all working to compare with the British Association unit, obtained results which finally left no doubt whatever as to the true relation. Dr. Werner Siemens's result found the mercury unit to be 0·9536 of the

British Association unit; Lord Rayleigh and
Mrs. Sidgwick found it 0·9542, which is an
exceedingly close agreement, being within $\frac{1}{10}$ per
cent. of the result of Dr. Werner Siemens. A
result differing by nearly one per cent. had been
obtained by Matthiessen and Hockin a good many
years before, when the precautions necessary to
reproduce the mercury standard with absolute
accuracy were not so well known as, in the course
of a few years after their work, they came to be
known. The final conclusion of Lord Rayleigh's
work was, that the Siemens mercury unit is 0·9413,
of what the Conference at Paris agreed to define
as the ohm; and that is the resistance measured
by 1,000,000,000 centimetres per second. I am
afraid that conveys a strange idea, but it is perfectly true as to the absolutely definite meaning
of resistance. I shall have occasion to refer to
the subject later, when I hope to explain this
mysterious velocity of 10^9 centimetres per second.
In the course of the thirty years from the time
when telegraphy began to demand definite
measurement, a great deal of accurate measure-

ment in terms of variously defined units of resistance had been made. Many sets of resistance coils had been produced by the Varley Brothers and other instrument makers, and many scientific investigators in laboratories had produced standards, and sets of resistance coils were made according to those standards ; but within the last twelve years all have merged into, either the Siemens, or the British Association, unit. The British Association unit, as I have said, was an attempt at absolute measurement, which succeeded in coming within 1·3 per cent. of the 10^9 aimed at. *Copies* of the British Association unit were *accurate to* $\frac{1}{10}$ *per cent.* The Siemens unit was founded on another idea, but it gave results no less definite and no less convenient for a great multitude of practical applications than did the somewhat nearer approach to a convenient absolute unit realised by the British Association Committee.

Gauss's principle of absolute measurement for magnetism and electricity is merely an extension of the astronomer's method of reckoning mass in terms of what we may call the universal-

gravitation unit of matter, and of the reckoning
of force adopted by astronomers, in common with
all workers in mathematical dynamics, according
to which the unit of force is that force, which
acting on unit of mass for unit of time, generates
a velocity equal to unit of velocity. The universal-
gravitation unit of mass is such a quantity of
matter, that if two quantities, each equal to it, be
placed at unit distance apart, the force between
them is unity.

The universal-gravitation method I refer to
for this reason. There is a terrestrial-gravitation
reckoning of force, according to the weight of
the unit of mass ; and after all, when we terrestrial
creatures take a mass in our hand and feel the
weight of it, it is a kind of measurement that we
cannot do away with. The kilogramme, or the
pound, or the ounce, is a thing we have to deal
with ; we have it in our hand, and we cannot help
using it to give us by its *heaviness* a reckoning
of force. A local gravitation unit of force means
the weight of a gramme in London, in Glasgow,
at the Equator, or anywhere else—and it is

a convenient unit; but the common mode of measuring force by reference to weight without reference to locality is not definite, because the weight of a gramme is different here from what it is at the Equator. The heaviness of a pound or a gramme is greater by a two-hundredth at either pole than at the Equator; or to give the exact figures, 0.00512. That is a difference of $\frac{1}{2}$ per cent., and if your accuracy is to be within a $\frac{1}{2}$ per cent., you cannot ignore the difference of the force of gravity in different places. But a vast number of measurements in engineering, and in the most ultra scientific work of scientific laboratories, does not aspire to so high a degree ot accuracy; and for all such work the local or terrestrial-gravitation unit suffices, without specifying what the particular place is—only that it is somewhere or other on the face of the earth. For instance, moduluses of rigidity, moduluses of rupture, breaking strains of material, are stated accurately enough for engineering purposes, in terms of a ton weight per square centimetre, or grammes weight per square centimetre, or any other such mode of reckoning; or if I had not vowed never to mention inches, I would say tons per square inch, which is common (perhaps too common) in engineering. All such measurements ignore the difference of gravity in different localities, except some more precise measurements in which an allowance for the force of gravity to reduce it to a standard of lat. 45° is made, or it is left to the person using the measurement to make the reduction. For all purposes, however, in which it would be desirable to apply a correction for the varying force of gravity in different places it is convenient to use Gauss's absolute unit, and not the terrestrial-gravitation unit of force. I may say in passing, that the mere idea, which lurked or was visibly manifested, according to the degree of understanding, in the old formula of elementary dynamics $F = m \dfrac{d\,v}{d\,t}$ was an immense step; and the realisation of that idea, the bringing of it into practical use, has contributed more than anything else I know to the intelligent treatment of dynamic problems and their application to both scientific and engineering matters. The system of absolute reckoning of force by Gauss cannot be too much commended, as a great and important practical improvement in the fundamental science of engineering and physics,—the science of dynamics. It consists simply in defining the unit of force as that force which, acting on a unit of mass for a unit of time, generates a velocity equal to the unit of velocity. It leaves the units of mass, length, and time to be assumed arbitrarily ; the gramme, the centimetre, and the mean solar second, for example, as in the now generally adopted " C. G. S." system.

But the universal-gravitation system of the dynamical astronomer defines the unit of mass in terms of the unit of length and the unit of force. I need not repeat the definition. Thus we have the interlocking of two definitions :—the unit of force defined in terms of the units of mass, length, and time ; the unit of mass defined in terms of the unit of force and the unit of length. It might seem as if we were proceeding in a vicious circle ; but the circle is not vicious,—the two definitions are logically and clearly inter-dependent. We have, as it were, two unknown quantities and two equations ; and the elimination of one of the unknown quantities from the two equations gives us the other explicitly. The two are mixed up in a somewhat embarrassing way in the primitive definitions, but when we disentangle them we arrive at the simple result, which I shall state presently, of independent definitions of the unit of mass and the unit of force, each in terms of units of length and time chosen arbitrarily.

Though the units of force and mass thus defined, are essentially implied in all the regular formulas of physical astronomy, from those most elementary ones, which appear in the treatment of the undisturbed elliptic motion, according to Newton's inferences from Kepler's laws, up to the most elaborate working out of the lunar, planetary, and cometary theories, and of the precession and nutation of the earth's axis; it has not been usual for physical astronomers to found any systematic numerical reckoning upon them, nor even, to choose arbitrarily and definitively any particular

units of length and time, on which to found the
units of force and mass. It is nevertheless interesting, not only in respect to the ultimate
philosophy of metrical systems, but also as full
of suggestions regarding the properties of matter,
to work out in detail the idea of founding the
measurements of mass and force on no other
foundation than the measurement of length and
time. In doing so we immediately find that the
square of an angular velocity is the proper measure
of density or mass per unit-volume; and that
the fourth power of a linear velocity is the proper
measure of a force. The first of these statements
is readily understood by referring to Clerk
Maxwell's suggestion, of taking the period of
revolution of a satellite revolving in a circle close
to the surface of a fixed globe of density equal
to the maximum density of water, as a fundamental unit for the reckoning of time. Modify
this by the independent adoption of a unit of time,
and we have in it the foundation of a measurement
of density, with the detail that the density of the
globe is equal to $3/(4\pi)$ of the square of this

H

satellite's angular velocity in radians[1] per second; that is, the square of the satellite's velocity, multiplied by 3 and divided by 4π, measures the density of the globe. It may be a hard idea to accept, but the harder it is the more it is worth thinking of, and the more instructive in regard to the properties of matter. There it is, explain it how you will, that the density of water, the density of brass, the mean density of the earth, is measured absolutely in terms of the square of an angular velocity. I do not know whether it is generally known, that to Fourier are due those dimensional equations that appear in the British Association's volume of reports, and in Clerk Maxwell's book, and in Everett's useful book *Units and Physical Constants*. The dimension for the reckoning of density is the square of an angular velocity on the universal-gravitation

---

[1] The radian is the unit in which angular velocity is expressed. It is an angle of $\left(\dfrac{180°}{\pi}\right)$ about $57°\!\cdot\!3$ (or more correctly $57°\!\cdot\!2958$). Thus an arm, or radius vector turning through an angle of about $57°\!\cdot\!3$ per second, is moving with unit angular velocity; or if the arm makes a complete circle in one second its angular velocity is $2\,\pi$.

absolute system, and is therefore T^{-2}. Equally puzzling and curious is a velocity to the fourth power for the reckoning of force, which we have next to consider.

The universal-gravitation reckoning of force, which we shall see is by the fourth power of a linear velocity, may be explained as follows. Find the velocity with which a particle of matter must be projected, to revolve in a circle round an equal particle fixed at such a distance from it as to attract it with a force equal to the given force. The fourth power of this velocity is the number which measures the force. Sixteen times the force will give double the velocity; eighty-one times the force will give three times the velocity, and so on.

Now if I were to say that the weight of that piece of chalk is the fourth power of twenty miles an hour, I should be considered fit, not for this place, but for a place where people who have lost their senses are taken care of. I suppose almost every one present would think it simple idiocy if I were to say that the weight of that

H 2

piece of chalk is the fourth power of seven or eight yards per hour; yet it would be perfectly good sense.

Think now of an infinitesimal satellite revolving round the earth—you ask, What is an infinitesimal satellite? To be "infinitesimal" for our present purpose, it must be very small in comparison with the earth, so as not to cause sensible motion by its reaction on the earth. Well, a 500-lb. shot is an infinitesimal satellite; though it is not, perhaps, infinitesimal in some of its aspects. There must be no resistance of the air, of course. Now fire it off with such a velocity that it will have a very flat trajectory, neither more nor less flat than the earth, and it will continue going round and round the earth. Find the velocity at which you must fire off the shot to make it go round the earth, and, if there is no resistance of the air, there is our infinitesimal satellite. These somewhat pedantic words are justified, because "infinitesimal satellite" is nine syllables to express three or four sentences; that is our justification.

The semi-period of an infinitesimal satellite revolving round the earth, close to its surface,[1] is equal to the semi-period of an ideal simple pendulum of length equal to the earth's radius, and having its weighted end infinitely near to the earth's surface ; and therefore, when reckoned in seconds, is approximately equal to the square root of the number of metres (6,370,000) in the earth's radius ; because the length of a seconds pendulum (or the pendulum whose semi-period is a second) is very approximately one metre. Thus we find 2,524 mean solar seconds for the semi-period of the satellite, and its angular velocity in radians per second is therefore $(\pi/2524 =)$ 0·001244 : hence the earth's mean density, reckoned on the universal-gravitation system, with the mean solar second for the unit of time, is $[(0·001244)^2 \times 3/(4\,\pi) =]$ 3·70 × 10^{-7} ; and, if we take (from Bailey's repetition of Cavendish's experiment),[2] the earth's mean density as 5·67

[1] Thomson and Tait's *Natural Philosophy*, 2nd edition, vol. i., part I., § 223.

[2] M. Cornu has criticised Bailey's method of reducing his observations, in respect to allowance for viscous diminution of the oscilla-

times the maximum density of water, we find 6.53×10^{-8} for the maximum density of water according to the universal-gravitation reckoning. To measure mass we must now introduce a unit of length, and if we take this as one centimetre, we find that, as the mass of a cubic centimetre of water at maximum density is very approximately equal to what is called a gramme, the universal-gravitation unit of matter is [$1/(6.53 \times 10^{-8})=$] 15.3×10^{6} grammes, or 15.3 French tons; hence the unit force on the universal-gravitation system is 15.6×10^{3} dynes; or 15.6 times the terrestrial weight of a kilogramme.

15.3 French tons, then (a French ton is 1.4 per cent. less than the British ton), is the universal-gravitation unit of matter. The time may come when the universal-gravitation system will be the system of reckoning; when 15.3 tons will be the unit of matter, and when the decimal subdivision of 15.3 French tons may be our metrical system,

tions of the torsion-rod. He has expressed the opinion that Bailey's result should, if calculated on thoroughly correct principles, have been in close agreement with his own, which was 5.55.

and grammes may be as much a thing of the past as grains are now.

There is something exceedingly interesting in seeing that we can practically found a metrical system on a unit of length and a unit of time. There is nothing new in it, since it has been known from the time of Newton, but it is still a subject full of fresh interest. The very thought of such a thing is full of many lessons in science that have scarcely yet been realised, especially as to the ultimate properties of matter. The gramme, it will be remembered, is founded on the properties of a certain body, namely, water; but here, without invoking any particular kind of matter, simply choosing a certain definite length marked on a measuring rod, and a unit of time (how obtained, we shall consider presently), we can take up a piece of matter, and tell, in any part of the universe, how to measure its mass in definite absolute units.

Think now of the two units on which this universal-gravitation metrical system depends: the unit of length and the unit of time. The unit of length is merely the length of a certain definite piece of brass, or other solid substance used for a measuring-rod, or the length between two marks upon it; it may be an inch, or a foot, or a yard, or a metre, or a centimetre—the principle is the same. The metre, it is true, was made originally as nearly as possible equal to the ten-millionth of the length of a certain quadrant of the earth, estimated as accurately as possible from the geodetic operations of MM. Méchain and Delambre in 1792, performed for the foundation of the metrical system. But this merely gave the original metre measure, and what is meant by the metre now is a length equal to it, or to some authentic copy which has been made from it as accurately as possible; and the one-hundredth part of the metre thus defined is the centimetre which we definitively adopt as the unit of length.

Thus our unit of length is independent of the earth, and is perfectly portable, so that the scientific traveller roaming over the universe carries his measuring-rod with him; and need think no more of the earth, so far as his measurement of space is concerned. But how about the mean solar second, in terms of which he measures his time? What of it, if he has left the earth for good; or if, even without leaving the earth, he carries on his scientific work on the earth through a few million years, in the course of which the period of the earth s rotation round its axis, and of its revolution round the sun, will both be very different from what they are now? If he takes a good watch or chronometer with him, well rated before he leaves the earth, it will serve his purpose as long as it lasts. What it does is merely to count the vibrations of a certain mass under the influence of a certain spring (the balance-wheel under the influence of the hair-spring). If, for any secular experiment he has in hand, he wishes to keep up a continuous reckoning of time, he must keep his watch always going, and not a vibration will be lost in the counting performed by the hands. But if he merely wishes to keep his unit of time, and to make quite sure that any number of million years hence, this shall be within one-tenth per cent. of its present value ; he should take a vibrator better arranged for permanence and for absolute accuracy, than the balance-wheel with its hair-spring of a watch or a chronometer. A steel tuning-fork, which has had its period of vibration determined for him, before he leaves the earth, by Professor Macleod, or by Lord Rayleigh, will serve his purpose. By measuring the period in terms of mean solar seconds, with the prongs up, and horizontal, and vertically down, he will be able to eliminate the slight effect of terrestrial gravity ; and he will have with him a time-standard that will give him the mean solar second, as accurately as his measuring-rod gives him the centimetre, in whatever part of the universe, and at whatever time, now or millions of years later, he has occasion to use his instruments.

I hope that you will not feel that I am abusing your good nature with an elaborate frivolity, when I ask you to think a little more of the unital equipment of our ideal traveller, on a scientific tour through the universe. For myself, what seems the shortest and surest way to reach the philosophy of measurement,—an understanding of what we mean by measurement, and which is essential to the intelligent practice of the mere art of measuring,—is to cut off all connection with the earth, and think what we must then do, to make measurements which shall be definitely comparable with those which we now actually make, in our terrestrial workshops and laboratories. Suppose, then, the traveller to have lost his watch and his tuning-fork and his measuring-rod ; but to have kept his scientific books, or at all events to have in his mind a full recollection and understanding of their contents : how is he to recover his centimetre, and his mean solar second ?

Let us consider the recovery of the centimetre first. Wherever he is let him make a piece of glass, like this which I hold in my hand, out of materials which he is sure to find, in whatever habitable region of the universe he may chance to be ; and let him with a diamond, or with a piece of hard steel, or with a piece of flint, engrave on it one thousand equidistant parallel lines, upon a space which may be about the breadth of his thumb, and which he may take as a temporary or provisional unit of length. He may help himself to engrave the glass by means of a screw cut in brass or steel, which he will easily make, though he has no tools, not even flint implements, to begin with. With a little time and perseverance he will make the requisite tools. Let him also make a temporary measuring-rod, and mark off equal divisions upon it, which may be of any convenient length, and need not have any relation to the definite provisional unit. Let him now make two candles, and light them and place them as you now see those on the table, at any convenient distance apart, measured on his measuring-rod. He holds the piece of ruled glass in his hand, close to his eye, as I hold this, and sees two rows of coloured spectrums, each with one of the candles in its centre. He turns the glass round till the two rows of spectrums are in the same line, and adjusts the parallelism of its plane, so as to make the distance from spectrum to spectrum a minimum. He moves backwards and forwards, as I do now, keeping his eye at equal distances from the two candles, until he sees each candle shooting up out of the yellow middle of a spectrum of the other candle, with no spectrum between the two candles. With this condition fulfilled, he measures the distance from the grating to the candles. Then, by the theory of diffraction, he has the proportion :—as the distance from the grating to the candles, is to the distance between the candles, so is the distance from centre to centre of the divisions on the glass, to the wavelength of yellow light. This, he remembers, is $5·892 \times 10^{-5}$ of a centimetre, and thus he finds the value in centimetres of his provisional unit.

[How easily this determination might be effected, supposing the grating once made, was illustrated by a rapid experiment performed in the course of the lecture ; without other apparatus than a little piece of glass with two hundred and fifty fine parallel lines engraved on it, two candles, and a measuring tape of unknown divisions of length (used only to measure the ratio between two distances). The result showed the distance from centre to centre of consecutive bars of the grating, to be thirty-two times the wave-length of yellow light. The breadth of the span on which the two hundred and fifty lines of the grating were ruled was thus measured as $(250 \times 32 \times 5 \cdot 892 \times 10^{-5} =)$ $0 \cdot 47136$ centimetre. According to the instrument-maker this space was said to be $0 \cdot 5$ of a centimetre.]

Thus you see, by this hurried experiment with this rough-and-ready apparatus, we have been able to measure a length to within a small percentage of accuracy. A few minutes longer spent upon the experiment, and using sodium flames behind fine slits instead of open candles blowing about in the air, with more careful measurement of the ratio of the distances, might easily have given a result within one-half per cent. of accuracy. Thus the cosmic traveller can easily recover his centimetre and his metre measure.

But how is our scientific traveller to recover his mean solar second, supposing he has lost his tuning-fork? He may think of the velocity of light, and go through Foucault's experiment.

That is a thing that can be done from the beginning, with nothing but cutting tools and pieces of metal to begin with. Let him get a piece of brass and make a wheel, and cut it to two thousand teeth. I do not know how many teeth Foucault used, but our traveller can go through the whole process, and set the wheel revolving at some uniform rate (not a known rate, because he has no reckoning of time); and he will tell what the velocity of the wheel is in terms of the velocity of light, which is known to be about 300,000 kilometres per second. If he is electrically minded, as this evening we are bound to suppose our scientific traveller to be, he will think of "v," or of an ohm. He may make a Siemens unit; that he can do, because he has his centimetre, and he finds mercury and glass everywhere. Then he goes through all that Lord Rayleigh and Mrs. Sidgwick have done. He will, with a temporary chronometer or vibrator, obtain a provisional reckoning of time, and he will go through the whole process of measuring the resistance of a Siemens unit in absolute measure, according to his provisional unit of time. His measurement gives him a velocity in, let us say, kilometres per this provisional unit of time, as the value of the Siemens unit in absolute measure. Then he knows from Lord Rayleigh and Mrs. Sidgwick, that the Siemens unit in absolute measure is 9,413 kilometres per mean solar second; and thus he finds the precise ratio of his provisional unit of time to the mean solar second.

Still, even though this method might be chosen as the readiest and most accurate, according to present knowledge of the fundamental data, for recovering the mean solar second, the method by "v" is too interesting and too instructive, in respect to elimination of the properties of matter from our ultimate metrical foundations, to be unconsidered. One very simple way of experimentally determining "v," is derivable from an important suggestion of Clark and Bright's paper referred to above. Take a Leyden jar, or other condenser of moderate capacity (for example, in electrostatic measure, about 1,000 centimetres) which must be accurately measured. Arrange a mechanism to charge it to an accurately known potential of moderate amount (for example, in electrostatic measure, about 10 C.G.S., which is about 3,000 volts), and discharge it through a galvanometer coil at frequent regular intervals (for example, ten times per any convenient unit of time). This will give an intermittent current of known average strength (in the example, 10^5 electrostatic C.G.S., or about 1/300,000 electro-magnetic C.G.S., or 1/30,000 of an ampère), which is to be measured in electro-magnetic units by an ordinary galvanometer. The number found by dividing the electrostatic reckoning of the current, by the experimentally found electro-magnetic reckoning of the same, is "v," in centimetres per the arbitrary unit of time, which the experimenter in search of the mean solar second has used in his electrostatic and electro-magnetic details. The unit of mass which he has chosen, also arbitrarily, disappears from the resulting ratio.

But there is another exceedingly interesting way—a way which, although I do not say it

I

is the most practical, has very great interest attached to it, as being the way of doing the thing in one process — that is the method of electrical oscillations.[1] I should certainly like to see how a person who has lost his standards, after having recovered his centimetre (which he certainly would do by the wave-length of light), would succeed in recovering his unit of time by the following method. Take a condenser—a very large Leyden jar ; electrify it, and connect the two poles through a conductor, arranged to have as large an electro-magnetic *quasi* inertia,[2]—electro-magnetic self-induction—as possible. The method is given in Clerk Maxwell's *Electricity and Mag-*

[1] See my Papers on "Transient Electric Currents," Glasgow Philosophical Society Proceedings, vol. III., Jan. 1853, and Philosophical Magazine, June, 1853 ; now constituting Article LXII. of my Reprint of "Mathematical and Physical Papers," vol. I., 1882.

[2] See on this subject my Paper "On the Mechanical Value of Distributions of Electricity, Magnetism, and Galvanism," read before the Glasgow Philosophical Society, January 1853, and published in their Proceedings (vol. iii.) for that date ; also article "Dynamical Relations of Magnetism," Nichol's "Cyclopædia of the Physical Sciences," 2nd edition, 1860. These two Papers, with additions of date July 1882, now constitute Article LXI. of my Reprint of "Mathematical and Physical Papers," vol. I., 1882.

netism (vol. ii. chap. xix.). It is too long to explain the details, but read the mathematical parts of Clerk Maxwell, read the British Association volume of Reports on Electrical Standards, and read Everett's *Units and Physical Constants ;* get these off by heart from the first word to the last, and you will learn with far less labour than by listening to me. Take a resistance coil of proper form for maximum electro-magnetic inertia,[1] and discharge the condenser through it ; or rather start the condenser to discharge through such a coil, and you will have a set of oscillations, following exactly the same law as the oscillations of the water-level in two cisterns, which, having initially had the free water-level in one higher than in the other, are suddenly connected by a U-tube. Imagine two such cisterns of water, connected by a U-tube with a stop-cock, and having the water higher in one cistern than in the other: now suddenly open the stop-cock, and the water-level will begin to fall in one cistern, and rise in the other. The inertia of the water, thus

[1] See Clerk Maxwell's Electricity and Magnetism, sect. 706.

I 2

made to flow through the connecting ∪-tube will cause it to flow on after it has come to its mean level in the two cisterns, and to rise to a higher level in the one in which it was previously higher, and to sink to a correspondingly lower level in the other. Thus the water-level in each cistern would alternately be above and below the mean free level: the range of motion becoming gradually diminished, in virtue of the viscosity of the water, until after a dozen or two of oscillations, the amplitude of each becomes so small that you cannot notice it. Precisely the same thing happens in the case of the discharge of a condenser through a resistance coil of large electro-magnetic inertia: the resistance of the copper wire being like the viscous influence which causes the oscillations of water to subside. If, in his investigations throughout the universe, our traveller could meet with a metal which is about a miliion times as conductive as copper, he would make this experiment with much greater ease; but it is practicable with copper. It is certain from the observations made by Feddersen, Schiller, and others, that a great number of oscillations can be observed, and that the period, or semi-period of oscillation, can be determined with considerable accuracy. If our scientific traveller wishes, by this beautiful experiment, to determine once for all his time reckoning, let him proceed thus. Let him take a coil, of which he knows the dimensions perfectly, having already gone through the preliminary process of measuring its electrical dimensions ; or if he cannot measure these with sufficient accuracy (and there is enormous difficulty in finding the electric dimensional qualities of a coil by measurement), let him do it partly by direct measurement of its length and of the linear dimensions of the figure into which it is wound, and partly by comparing it electro-magnetically with other coils. By an elaborate investigation he can find the electro-magnetic inertia of the coil in terms of his centimetre. And here, again, there is a curious kind of puzzle and apparent incongruity, when I say that the electro-magnetic inertia equivalent of a coil is a length, and is measured as a numeric of centimetres. Let him make a condenser, and

by building it up from small to large, let him learn
the capacity of it in electrostatic measure. Let
him begin with two plates or cylinders, or a sphere
enclosed within a concentric sphere, and go on
multiplying till he gets a capacious enough condenser of which he knows, in electrostatic measure,
the electrostatic capacity. This, again, is a line.
Now let him take the rectangle of those two lines,
and construct the equivalent square—let him,
geometrically or arithmetically, take the square
root of the product of the two lines—and let him
observe the period of electric oscillation that I
have spoken of. Let him imagine the hand of
a watch, going once round in the observed period.
He has good magnetic eyes, and he sees the
electro-magnetic oscillation, or he has appliances
by which he can test it : the thing has been done.
He sets in motion a little piece of wheel-work, with
a hand going once round in the period of the
oscillation. Now for a moment let him imagine
that hand to be equal in length to the square root
of the product of those two lines—several million
centimetres, or several thousand kilometres, if

the coil and condenser are of dimensions convenient for the actual experiment, as we terrestrials
might do it. The velocity of the end of that hand
is "*v.*" There he has this wonderful quantity
"*v.*" He has a hand going round in a certain
time, and he knows that if that hand be of the
calculated length, the velocity of the end of it
is "*v.*" This is interesting and instructive, and
though I do not for certain know that it is
very practicable, it is still, I believe, sufficiently
so to be worth thinking of. I think it will be
one of the ways of determining this marvellous
quantity "*v.*"

It is to be hoped that before long "*v*" will be
known, in centimetres per mean solar second,
within 1 10 per cent. At present it is only known
that it does not *probably* differ 3 per cent. from
2·9 × 10¹⁰ centimetres per mean solar second.
When it is known with satisfactory accuracy, an
experimenter provided with a centimetre measure
may, anywhere in the universe, rate his experimental chronometer to mean solar seconds, by
the mere electrostatic and electro-magnetic opera-

tions described above, without any reference to the sun or other natural chronometer.

I have tried your patience, I fear, too long, but I have now only reached the threshold of my subject. We now must commence the consideration of electrical units of measurement. I need not go round defining quantities electrostatically and electro-magnetically; you will find it all in Everett, and in the British Association volume of collected Reports by the first Committee on Electric Measurement. It is not for me to tell you of an ohm, a volt, a micro-farad, and so on; but there are two or three points that I should like to notice, and one is, the limitation of the so-called practical system. The absolute system goes from beginning to end in a perfectly consistent manner, with the initial conditions carried out all through; one of which, in the electromagnetic system, is that the electromotive force produced by the motion at unit speed, across the lines of force of a field of unit intensity, of a unit length of conductor, is unity. That you must carry out if the system is to be complete and consistent, and the dimensions of all your instruments and apparatus must all be all reckoned uniformly in terms of the unit of length adopted in the absolute definition. The ohm is 1,000,000,000 centimetres, or 10,000 kilometres, per second. If we are to make the ohm an absolute electromagnetic unit with the second as the unit of time, we must take the earth's quadrant as the unit of length. If we take that consistently throughout, we need never leave this particular system and we need have nothing to do with C.G.S. We should have the Q.G.S. system pure and simple! But it would be obviously inconvenient to measure the dimensions of instruments, the diameters of wheels, and the gauges of wire in submultiples of the earth's quadrant. Imagine the horror of a practical workman, on hearing a scientific person say to him, "Give me a wire 1/100,000 of an earth-quadrant long, and 1/10,000,000,000 in diameter." Now wherein does the so-called practical system differ from the absolute system, and why is it not to be as logical and complete as the absolute system? We would never leave the absolute system, if it gave us in all cases convenient numbers; and it does give us convenient numbers for the measurement of a current, its unit being ten times the "ampère" of the practical system. The unit of resistance in C.G.S., however, is too small, so is the unit of electromotive force. To get convenient numbers, we give names to certain multiples of units, that is all; and we use these multiples just as long as it is convenient, and not any longer. That is my idea of the practical system—to use it for convenience and as long as it is convenient; the moment it ceases to be convenient, to throw it overboard and take C.G.S. pure and simple. The Conference at Paris decided upon the practical system, by adopting the units which are now so familiar, the ohm, the volt (taken from the British Association recommendation), and the ampère. The coulomb was also added, and it was most satisfactory to get old Coulomb's name in—one of the fathers of electrical science. Then the watt was added by Sir W. Siemens, and it has been generally accepted, and has proved exceedingly convenient. But when you go farther with the practical system, and take anything that involves a magnetic pole or a magnetic field, you get lost in the trouble of adopting the earth's quadrant as unit of length, and deviation from C.G.S. ceases to be convenient. Return then to C.G.S. pure and simple.

I spoke of the resistance of an ohm being measured in terms of a velocity. I should like to explain this in a few words. Imagine a mouse-mill set with its axis vertical. Put a pair of brushes at the tops and bottoms of the bars ; put the brushes in the magnetic north and south plane through the axis, and set the mouse-mill to spin at any rate you please. Take a galvanometer like a tangent galvanometer, but with only an arc of wire equal in length to the radius—an arc subtending an angle equal to about $57°.3$— having its ends on the same level, whether above or below the level of the needle, and electrodes perpendicular to the plane of the arc connected with the brushes. The mouse-mill must be placed so far from the galvanometer, as not sensibly to influence it by electro-magnetic force. Now take the galvanometer and turn the mouse-mill; let the length of each bar of the mouse-mill be a centimetre; but that would be a flea-mill rather than a mouse-mill—say, let each bar be 100 centimetres; turn the mouse-mill round fast enough to cause your galvanometer to be deflected 45°. Then one hundred times the velocity of the bars is equal to the resistance in the circuit. Double resistance requires double velocity; half resistance requires half velocity to give the prescribed 45° deflection. There, then, is the rationale of 10,000 kilometres per second, or 1,000,000,000 centimetres per second being the measure of resistance. While we thus measure resistance in electromagnetic measure by velocity, we measure a conductivity in electrostatics by a velocity. I have given a very simple explanation of this also in a statement quoted by Sir William Siemens in his presidential address to the British Association at Southampton in 1882. The velocity at which the surface of a globe must shrink towards the centre, to keep its potential constant, when it is connected

to the earth by a wet thread, measures the conducting power of that wet thread. Double conducting power will require double velocity of
shrinkage, that is, the globe must shrink twice
as fast not to lose its potential. With a very
long semi-dry thread the globe may shrink
slowly. Suppose we have a globe insulated in
the air of this room for electrical experiment, and
connected with the ground by a silk thread. If
you have an electrometer to show the potential,
you will see it gradually sink. You might
imagine that dust in the air would carry off
electricity, but in truth practically the sole loss
is by this semi-dry silk thread. When you see
the potential sinking, imagine you see the globe
shrinking slowly, so as to keep its potential
constant, while it is gradually losing its electric
charge little by little : the velocity with which the
surface must shrink towards the centre to keep
the potential constant measures the conducting
power of the silk thread in electrostatic measure.
Thus we learn how it is a velocity which measures in electrostatic measure the conducting

power of a certain thread or wire. But, as we have seen in electro-magnetic measure, the resistance of the same thread or wire is measured by another velocity. The mysterious quantity "v" is the square root of the product of the two velocities. Or it is the one velocity which measures in electro-magnetic measure the resistance, and in electro-static measure the conductivity, of one and the same conductor; which must be of about 29 ohms resistance, because experiment has proved "v" to be not very different from 290,000 kilometres per second.

I have spoken to you of how much we owe to Sir Charles Bright and Mr. Latimer-Clark for the suggestion of names. How much we owe for the possession of names, is best illustrated by how much we lose—how great a disadvantage we are put to—in cases in which we have not names. We want a name for the reciprocal of resistance. We have the name "conductivity," but we want a name for the unit of conductivity. I made a box of resistance coils thirty years ago, and another fifteen years ago, for the measurement of conductivity, and they both languished for the want of a name. My own pupils will go on using the resistance box in ohms, rather than the conductivity box, because in using the latter it is so puzzling to say " The resistance is the reciprocal of the sum of the reciprocals of these resistances." It is the conductivity that you want to measure, but the idea is too puzzling ; and yet for some cases the conductivity system is immensely superior in accuracy and convenience to that by adding resistances in series. For the reciprocal of an ohm in the measurement of resisting power—for the unit reckoning of conductivity which will agree with the ohm—it is suggested to take a phonograph and turn it backwards, and see what it will make of the word "ohm." I admire the suggestion, and I wish some one would take the responsibility of adopting it ; we should then have *mho* boxes of coils at once in general use. With respect to electric light, what is it we want to measure by the current galvanometer ? We have a potential galvanometer, and we have a current galvanometer. Everybody knows what we want to measure with the potential galvanometer. The servant in every house that is lighted electrically knows about potentials; and if in reading the galvanometer he sees it is down to eighty volts he knows that something is wrong, and will at once go to the engine-room and cause eighty-four volts to be supplied; supposing, for example (as in the case of my own house, temporarily, until I can get two-hundred-volt lamps), that the proper potential is eighty-four volts. But in the current galvanometer there are so many divisions indicating, it may be, the number of amperes in the current. But after all, what do we want besides a knowledge of the potential? It is the sum of the reciprocals of the resistances in the circuit. In the multiple-arc system each fresh lamp lighted adds a conductivity. In a circuit of Edison or Swan hundred-volt lamps, in each of which you have a current of 0·7 of an ampere, and therefore a resistance of 143 ohms, how convenient it would be, in putting on a lamp—adding a certain conductivity—if we could say we add a *mho*, or a fraction of a *mho*, as the case may be. I do not say that *mho* is the word to be used, but I wish it could be accepted, so that we might have it at once in general use. We shall have a word for it when we have the thing, or rather, I should say, we shall have the thing when we have the word. The Appendix to the 1862 Report of the first British Association Committee on Electric Measurements contains a description of a " Resistance Measurer " invented by Sir William Siemens, and of a " Modification of Siemens' Resistance Measurer," by Professor Fleeming Jenkin. This instrument gives directly the resistance of a conductor, by means of an instrumental adjustment, bringing a magnetic needle to a zero position for each observation. In the original Siemens instrument the adjustment is a shifting of two coils by translational motion, and the conductivity is read on a scale of equal divisions adapted, by means of a curve determined by experiment, to give a reading of the required resistance. In Jenkin's modification the mechanical arrangement is much simplified by the adoption of a different electro-magnetic combination ; and

K

the required resistance is given by the tangent of the angle through which the coils must be turned to bring the needle to zero. A similar instrument to give conductivity by a simple reading, without any adjusting or "setting" for each observation, is easily made. I made such an instrument in 1858, being simply a galvanometer with controlling resistance coils instead of controlling magnet.[1] Such an instrument at once gives conductivity, and you want a name (suppose you adopt *mho*) for the unit of conductivity, and call the instrument a mhometer. The rule for resistances in series would be, the sum of the reciprocals of mhos is equal to the number of ohms; and for conductivities in parallels, the sum of the reciprocals of ohms is equal to the number of mhos. The number of mhos, or of millimhos, will then measure the number of lamps in the circuit. The domestic incandescent lamp of the early future ought to be, and we hope will

---

[1] This instrument is represented in Fig. 6 of my patent No. 329 of 1858 for "Improvements in Testing and Working Electric Telegraphs."

be, a one-millimho lamp, to give a ten- or twelve-
candle light with the Board of Trade regulated
200 volts of potential. Thus the lamp-galvano-
meter, or lamp-counter, may have its scale divided
to one millimho to the division, and the number
read on its scale at any time will be simply the
number of lamps lighted at the time.[1] The instrument will also have the great advantage of
being steady, notwithstanding the variations of
the engine. A potential instrument on an electric-
light circuit at best is always somewhat variable,
because the potential varies a good deal—within
one or two per cent. perhaps; but the resistance
in the lamps varies exceedingly little. The mhometer will in these circumstances be an absolutely
steady instrument; you will not see it quiver, even
though the engine is irregular. The potential
galvanometer will show you how much unsteadiness
there is to be complained of or to be corrected.

---

[1] [Note of December 8, 1887. A form of magnetostatic tangent
galvanometer, which I have recently brought out for practical use,
serves the same purpose. It is of simpler construction and more
convenient form than the mhometer referred to in the text.—W. T.]

Lastly, as to the objects to be aimed at in respect to the use of this great system of units. Nothing can be much more satisfactory than is the measurement of somewhat large resistances, as we have it habitually as present ; but if we want a better method for low resistances, we will be helped very much by the use of the mho boxes of conductivities which I have indicated. The great thing we want now in the way of practical electric measurement is a good standard of electromotive force. That was the chief object of a recent British Association Committee, but it has not yet been satisfactorily attained for practical purposes. Standard cells serve for the purpose to some extent, but we want something better, something of the nature of an electro-dynamometer, to give a good steady idiostatic potential gauge, by which the constant of any electrometer or ordinary galvanometer may be easily and accurately tested. That is an object to be sought; there are plenty of ways of obtaining it, and I hope, before another year has passed, to see it realised in many ways, certainly in one way.

As to the science of electricity, the great want in the way of measurement just now is the accurate measurement of "*v*," the ratio between the electrostatic and the electro-magnetic units; and I hope that scientific investigators will take the matter up, and give to it an accuracy like that which Lord Rayleigh has given to the measurement of the ohm.

A most interesting point remains. It is Joule's work, reported on by the British Association Committee :—see volume of Reports on Electrical Standards (Spon, 1871), p. 138. It was only in my preparation for this lecture that I came upon it, and put the figures definitely together. Joule, with a modesty characteristic of the man, and with a magical accuracy characteristic of his work, made, at the request of the British Association, an investigation of the heating effect of a current measured in a definite way, according to the measure of resistance of the British Association ohm, supposed then to be 10^9 C. G. S. units of resistance; and he himself considered that the electrical measurement which he then made was more accurate than his old frictional measurement of the mechanical equivalent of the thermal unit could be. The result obtained, assuming the British Association ohm to be absolutely correct, gave the mechanical equivalent as 782·2 foot-pounds, instead of 772 which he had made it before, and he expressed himself willing to make a new determination of it by the frictional method. But now let us put ourselves in the position of 1867, the date of this report, with these competing determinations of the ohm : the one obtained by the British Association method of spinning coils, and the other by Joule's electro-thermal method ; taking the dynamical value of the thermal unit, as given by his frictional method. Supposing that this electro-thermal method was right, then what we are to infer is not that the result is the mechanical equivalent, but that the British Association unit was not 10^9, as it was supposed to be, but $10^9 \times 0.98697$. Thus this experiment was virtually Joule's determination of the resistance of the British Association ohm in absolute measure. Lord Rayleigh's determination is $10^9 \times 0.98677$, a difference of 2 in the 4th place, within about 1/50 per cent. There is perfect magic in the accuracy of Joule's work: it is not a matter of chance. I think, between Joule, Lord Rayleigh, Mrs. Sidgwick, and others, we cannot have much doubt now, what is the absolute value of the Siemens unit, or of the British Association unit. I advise everybody to take the Rayleigh ohm unit, instead of the British Association ohm. I have begun to do so, and I mark everything R. O. You may have everything in the British Association unit, but reduce, if you please, to Rayleigh ohms by the reducing factor 0.98677. Volts must be reduced in the same ratio. The old estimate which I made in 1851 from Joule's experiment, for the absolute electromotive force of a standard Daniell cell, was 1.07 volts; and after thinking it was 1.078 for ten years, because of the British Association unit, we come back to correct it, and find it is 1.07. So much for the volt. But we want far more accurate instruments and methods connected with other parts of electric measurement, especially electromotive force and capacity, electro-static, or electro-magnetic, with the comparing number "*v*." These are the things we want to advance and perfect, in order to give a satisfactorily scientific character to this great system of absolute measurement, of which I have endeavoured to trace and explain the origin.

THE SORTING DEMON OF MAXWELL.

[*Abstract of a Friday evening Lecture before the Royal Institution of Great Britain,* February 28, 1879 (*Proc. R. I.* vol. ix. p. 113).]

THE word "demon," which originally in Greek meant a supernatural being, has never been properly used as signifying a real or ideal personification of malignity.

Clerk Maxwell's "demon" is a creature of imagination having certain perfectly well defined powers of action, purely mechanical in their character, invented to help us to understand the " Dissipation of Energy " in nature.

He is a being with no preternatural qualities, and differs from real living animals only in extreme smallness and agility. He can at pleasure stop, or strike, or push, or pull any single atom of matter, and so moderate its natural course of motion. Endowed ideally with arms and hands and fingers—two hands and ten fingers suffice— he can do as much for atoms as a pianoforte player can do for the keys of the piano—just a little more, he can push or pull each atom *in any direction.*

He cannot create or annul energy; but just as a living animal does, he can store up limited quantities of energy, and reproduce them at will. By operating selectively on individual atoms he can reverse the natural dissipation of energy, can cause one half of a closed jar of air, or of a bar of iron, to become glowing hot and the other ice-cold; can direct the energy of the moving molecules of a basin of water to throw the water up to a height and leave it there proportionately cooled (1 deg. Fahrenheit for 772 ft. of ascent); can " sort " the molecules in a solution of salt or in a mixture of two gases, so as to reverse the natural process of diffusion, and produce con- centration of the solution in one portion of the water, leaving pure water in the remainder of the space occupied ; or, in the other case, separate the gases into different parts of the containing vessel.

"Dissipation of Energy" follows in nature from the fortuitous concourse of atoms. The lost motivity is essentially not restorable otherwise than by an agency dealing with individual atoms ; and the mode of dealing with the atoms to restore motivity is essentially a process of assortment, sending this way all of one kind or class, that way all of another kind or class.

The classification, according to which the ideal demon is to sort them, may be according to the essential character of the atom ; for instance, all atoms of hydrogen to be let go to the left, or stopped from crossing to the right, across an ideal boundary ; or it may be according to the velocity each atom chances to have when it approaches the boundary : if greater than a certain stated amount, it is to go to the right ; if less, to the left. This latter rule of assortment, carried into execution by the demon, disequalises temperature, and undoes the natural diffusion of heat; the former undoes the natural diffusion of matter.

By a combination of the two processes, the demon can decompose water or carbonic acid, first raising a portion of the compound to dissociational temperature (that is, temperature so high that collisions shatter the compound molecules to atoms), and then sending the oxygen atoms this way, and the hydrogen or carbon atoms that way; or he may effect decomposition against chemical affinity otherwise, thus:—Let him take in a small store of energy by resisting the mutual approach of two compound molecules, letting them press as it were on his two hands, and store up energy as in a bent spring; then let him apply the two hands between the oxygen and the double hydrogen constituents of a compound molecule of vapour of water, and tear them asunder. He may repeat this process until a considerable proportion of the whole number of compound molecules in a given quantity of vapour of water, given in a fixed closed vessel, are separated into oxygen and hydrogen at the expense of energy taken from translational motions. The motivity (or energy for motive power) in the explosive mixture of oxygen and hydrogen of the one case, and the separated mutual combustibles, carbon and oxygen, of the other case, thus obtained, is a transformation of the energy found in the substance in the form of kinetic energy of the thermal motions of the compound molecules. Essentially different is the decomposition of carbonic acid and water in the natural growth of plants, the resulting motivity of which is taken from the undulations of light or radiant heat, emanating from the intensely hot matter of the sun.

The conception of the "sorting demon" is merely mechanical, and is of great value in purely physical science. It was not invented to help us to deal with questions regarding the influence of life and of mind on the motions of matter, questions essentially beyond the range of mere dynamics.

The discourse was illustrated by a series of experiments.

ELASTICITY VIEWED AS POSSIBLY A MODE OF MOTION.

[*Abstract of a Friday evening Lecture before the Royal Institution of Great Britain*, March 4, 1881 (*Proc. R. I.* vol. ix. p. 520).]

WITH reference to the title of his discourse the speaker said: "The mere title of Dr. Tyndall's beautiful book, *Heat, a Mode of Motion*, is a lesson of truth which has manifested far and wide through the world one of the greatest discoveries of modern philosophy. I have always admired it; I have long coveted it for Elasticity; and now, by kind permission of its inventor, I have borrowed it for this evening's discourse."

"A century and a half ago Daniel Bernouilli shadowed forth the kinetic theory of the elasticity of gases, which has been accepted as truth by

Joule, splendidly developed by Clausius and Maxwell, raised from statistics of the swayings of a crowd to observation and measurement of the free path of an individual atom in Tait and Dewar's explanation of Crookes' grand discovery of the radiometer, and in the vivid realisation of the old Lucretian torrents with which Crookes himself has followed up their explanation of his own earlier experiments ; by which, less than two hundred years after its first discovery by Robert Boyle, 'the Spring of Air' is ascertained to be a mere statistical resultant of myriads of molecular collisions."

"But the molecules or atoms must have elasticity, and *this* elasticity must be explained by motion before the uncertain sound given forth in the title of the discourse, 'Elasticity viewed as possibly a Mode of Motion,' can be raised to the glorious certainty of 'Heat, a Mode of Motion.'"

The speaker referred to spinning-tops, the child's rolling hoop, and the bicycle in rapid motion as cases of stiff, elastic-like firmness produced by motion; and showed experiments with gyrostats in which upright positions, utterly unstable without rotation, were maintained with a firmness and strength and elasticity such as might be by bands of steel. A flexible endless chain seemed rigid when caused to run rapidly round a pulley, and when caused to jump off the pulley, and let fall to the floor, stood stifflly upright for a time till its motion was lost by impact and friction of its links on the floor. A limp disc of indiarubber caused to rotate rapidly seemed to acquire the stiffness of a gigantic Rubens hat-brim. A little wooden ball, which when thrust down under still water jumped up again in a moment, remained down as if embedded in jelly when the water was caused to rotate rapidly, and sprang back, as if the water had elasticity like that of jelly, when it was struck by a stiff wire pushed down through the centre of the cork by which the glass vessel containing the water was filled. Lastly, large smoke rings discharged from a circular or elliptic aperture in a box were rendered visible, by aid of the electric light, in their progress through the air of the theatre. Each ring was circular, and its motion was steady when the aperture from which it proceeded was circular, and when it was not disturbed by another ring. When one ring was sent obliquely after another the collision or approach to collision sent the two away in greatly changed directions, and each vibrating seemingly like an indiarubber band. When the aperture was elliptic each undisturbed ring was seen to be in a state of regular vibration from the beginning, and to continue so throughout its course across the lecture-room. Here, then, in water and air was elasticity as of an elastic solid, developed by mere motion. May not the elasticity of every ultimate atom of matter be thus explained ? But this kinetic theory of matter is a dream, and can be nothing else, until it can explain chemical affinity, electricity, magnetism, gravitation, and the inertia of masses (that is, crowds) of vortices.

Le Sage's theory might give an explanation of gravity and of its relation *to inertia of masses*, on the vortex theory, were it not for the

L

essential æolotropy of crystals, and the seemingly perfect isotropy of gravity. No finger-post
pointing towards a way that can possibly lead
to a surmounting of this difficulty, or a turning
of its flank, has been discovered, or imagined as
discoverable. Belief that no other theory of
matter is possible is the only ground for anticipating that there is in store for the world another
beautiful book to be called *Elasticity, a Mode of
Motion.*

THE SIZE OF ATOMS.

[*Friday evening Lecture before the Royal Institution of Grea
Britain*, February 3, 1883 (*Proc. R. I.* vol. x. p. 185).]

FOUR lines of argument founded on observation
have led to the conclusion that atoms or molecules
are not inconceivably, not immeasurably small.
I use the words "inconceivably" and "immeasurably" advisedly. That which is measurable is not
inconceivable, and therefore the two words put
together constitute a tautology. We leave in-
conceivableness in fact to metaphysicians. Nothing
that we can measure is inconceivably large or
inconceivably small in physical science. It may
be difficult to understand the numbers expressing
the magnitude, but whether it be very large or
very small there is nothing inconceivable in the
nature of the thing because of its greatness or

L 2

smallness, or in our views and appreciation and numerical expression of the magnitude. The general results of the four lines of reasoning to which I have referred, founded respectively on the undulatory theory of light, on the phenomena of contact electricity, on capillary attraction, and on the kinetic theory of gases, agree in showing that the atoms or molecules of ordinary matter must be something like the 1/10,000,000th or from the 1/10,000,000th to the 1/100,000,000th of a centimetre in diameter. I speak somewhat vaguely, and I do so not inadvertently, when I speak of atoms and molecules. I must ask the chemists to forgive me if I even abuse the words and apply a misnomer occasionally. The chemists do not know what is to be the atom; for instance, whether hydrogen gas is to consist of two pieces of matter in union constituting one molecule, and these molecules flying about; or whether single molecules, each indivisible, or at all events undivided in chemical action, constitute the structure. I shall not go into any such questions at all, but merely take the broad view that matter, although we may conceive it to be infinitely divisible, is not infinitely divisible without decomposition. Just as a building of brick may be divided into parts, into a part containing 1000 bricks, and another part containing 2500 bricks, and those parts viewed largely may be said to be similar or homogeneous ; but if you divide the matter of a brick building into spaces of nine inches thick, and then think of subdividing it farther, you find you have come to something which is atomic, that is, indivisible without destroying the elements of the structure. The question of the molecular structure of a building does not necessarily involve the questions, Can a brick be divided into parts? and Can those parts be divided into much smaller parts? and so on. It used to be a favourite subject for metaphysical argument amongst the schoolmen whether matter is infinitely divisible, or whether *space* is infinitely divisible, which some maintained ; whilst others maintained that *matter* only is not infinitely divisible, and demonstrated that there is nothing inconceivable in the infinite subdivision of space. Why, even time was divided into moments (time-atoms!), and the idea of continuity of time was involved in a halo of argument, and metaphysical—I will not say absurdity—but metaphysical word-fencing, which was no doubt very amusing for want of a more instructive subject of study. There is in sober earnest this very important thing to be attended to, however, that in chronometry, as in geometry, we have absolute continuity, and it is simply an inconceivable absurdity to suppose a limit to smallness whether of time or of space. But on the other hand, whether we can divide a piece of glass into pieces smaller than the 1/100,000th of a centimetre in diameter, and so on without breaking it up, and making it cease to have the properties of glass, just as a brick has not the property of a brick wall, is a very practical question, and a question which we are quite disposed to enter upon.

I wish in the beginning to beg you not to run away from the subject by thinking of the exceeding smallness of atoms. Atoms are not so exceedingly small after all. The four lines of argument I have referred to make it perfectly certain that the molecules which constitute the air we breathe are not very much smaller, if smaller at all, than 1/10,000,000th of a centimetre in diameter. I was told by a friend just five minutes ago that if I gave you results in centimetres you would not understand me. I do not admit this calumny on the Royal Institution of Great Britain; no doubt many of you as Englishmen are more familiar with the unhappy British inch; but you all surely understand the centimetre; at all events it was taught till a few years ago in the primary national schools. Look at that diagram (Fig. 33), as I want you all to understand an inch, a centimetre, a millimetre, the 1/10th of a millimetre, the 1/100th of a millimetre, the 1/1000th of a millimetre, and the 1/1,000,000th of a millimetre. The diagram on the wall represents the metre; below that the yard; next the decimetre, and a circle of a decimetre diameter, the centimetre, and

a circle of a centimetre, and the millimetre, which
is 1/10th of a centimetre (or in round numbers
1/40th of an inch), and a circle of a millimetre.
(For convenience the woodcut Fig. 33, representing
the diagram in question, shews the relative dimension of the centimetre and millimetre only.) We

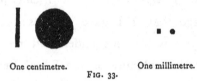

One centimetre. One millimetre.
Fig. 33.

will adhere however to one simple system, for it
is only because we are in England that the yard
and inch are put before you at all, among the
metres and centimetres. You see on the diagram
then the metre, the centimetre, the millimetre, with
circles of the same diameter. Somebody tells me
the millimetre is not there. I cannot see it, but
it certainly is there, and a circle whose diameter
is a millimetre, both accurately painted in black.
I say there is a millimetre, and you cannot see
it. And now imagine *there* is 1/10th of a
millimetre, and *there* 1/100th of a millimetre,

and *there* 1/1000th of a millimetre, and *there* a round atom of oxygen 1/1,000,000th of a milli-metre in diameter. You see them all.

Now we must have a practical means of mea-suring, and optics supply us with it, for thou-sandths of a millimetre. One of our temporary standards of measurement shall be the wave-length of light; but the wave-length is a very indefinite measurement, because there are different wave-lengths for different colours of light, visible and invisible, in the ratio of 1 to 16. We have, as it were—borrowing an analogy from sound—four octaves of light that we know of. How far the range in reality extends, above and below the range hitherto measured, we cannot even guess in the present state of science. The table before you (Table I.) gives you an idea of magnitudes of length, and again of small intervals of time. In the column on the left you have the wave-length of light in fractions of a centimetre; the unit in which these numbers is measured being the 1/100,000th (or 10^{-5}) of a centimetre. We have then, of visible light, wave-lengths from

$7\frac{1}{2}$ to 4 nearly, or 3·9. You may say then, roundly, that for the wave-lengths of visible light, which alone is what is represented on that table, we

TABLE I.—DATA FOR VISIBLE LIGHT

Line of Spectrum.	Wave-length in Centimetres.	Wave Frequency, or Number of Periods per Second.
A	$7·604 \times 10^{-5}$	$395·0 \times 10^{12}$
B	6·867 ,,	437·3 ,,
C	6·562 ,,	457·7 ,,
D_1	5·895 ,,	509·7 ,,
D_2	5·889 ,,	
E	5·269 ,,	570·0 ,,
b	5·183 ,,	
F	4·861 ,,	617·9 ,,
G	4·307 ,,	697·3 ,,
H_1	3·967 ,,	756·9 ,,
H_2	3·933 ,,	763·6 ,,

have wave-lengths of from 4 to 8 on our scale of 1/100,000th of a centimetre. The 8 is invisible radiation a little below the red end of the spectrum. The lowest, marked by Fraunhofer with the letter A, has for wave-length $7\frac{1}{2}$/100,000ths of a centimetre. On the model before you I will now show you what is meant by a "wave-length;" it is not length along the crest, such as we sometimes see well marked in a wave of the sea breaking on a long straight beach; it is distance from crest to crest of the waves. [This was illustrated by a large number of horizontal rods of wood connected together and suspended bifilarly by two threads in the centre hanging from the ceiling;[1] on moving the lowermost rod, a wave was propagated up the series.] Imagine the ends of those rods to represent particles. The rods themselves let us suppose to be invisible, and merely their ends visible, to represent the particles acting upon one another mutually with elastic force, as if of indiarubber bands, or steel spiral springs, or jelly, or elastic material of some kind. They do act on one another in this model through the

[1] The details of this bifilar suspension need not be minutely described, as the new form, with a single steel pianoforte wire to give the required mutual forces, described below and represented in Fig. 34, is better and is more easily made.

central mounting. Here again is another model
illustrating waves (Fig. 34).[1] The white circles on

[1] This apparatus, which is represented in the woodcut, Fig. 34, is
of the following dimensions and construction. The series of equal
and similar bars (B) of which the ends represent molecules of the
medium, and the pendulum bar (P), which performs the part of
exciter of vibrations, or of kinetic store of vibrational energy, are
pieces of wood each 50 centimetres long, 3 centimetres broad, and
1·5 centimetres thick. The suspending wire is steel pianoforte wire
No. 22 B. W. G. (·07 of a cm. diameter), and the bars are secured
to it in the following manner. Three brass pins of about 4 of a
centimetre diameter are fitted loosely in each bar in the position
indicated ; *i.e.* forming the corners of an isosceles triangular figure,
with its base parallel to the line of the suspending wire, and about
1 mm. to one side of it. The suspending wire, which is laid in
grooves cut in the pins, is passed under the upper pin, outside the
pin at the apex of the triangle, over the upper side of the lower pin,
and thence down to the next bar. The upper end of this wire is
secured by being taken through a hole in the supporting beam and
several turns of it put round a pin placed on one side of the hole, as
indicated in the diagram. To each end of the pendulum bar is made
fast a steel spiral spring, as shown ; the upper ends of these springs
being secured to short cords which pass up through holes in the
supporting beam, and are fastened by two or three turns taken
round the pins. These steel springs serve as potential stores of
vibrational energy alternating in each vibration with the kinetic
store constituted by the pendulum bar. The ends of the vibrating
bars (B) are loaded with masses of lead attached to them. The
much larger masses of lead seen on the pendulum bar, which are
adjustable to different positions on the bar, are, in the diagram,
shown at the smallest distance apart. The lowermost bar carries
two vanes of tin projecting downwards, which dip into viscous

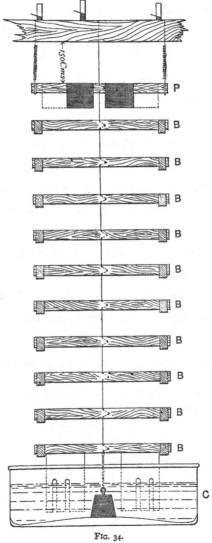

FIG. 34.

the wooden rods represent pieces of matter—I
will not say molecules at present, though we
shall deal with them as molecules afterwards.
Light consists of vibrations transverse to the
line of propagation, just as in the models before
you.

Now in that beautiful experiment well known
as Newton's rings we have at once a measure of
wave-length in the distance between two pieces
of glass to give any particular tint of colour.
The wave-length, you see, is the distance from
crest to crest of the waves travelling up the long
model when I commence giving a simple harmonic oscillation to the lowest bar. I have here
a convex lens of very long focus, and a piece of
plate glass with its back blackened. When I press
the convex lens against the piece of plate glass

liquid (treacle diluted with water) contained in the vessel (C). A
heavy weight resting on the bottom of this vessel, and connected to
the lower end of the suspending wire by a stretched indiarubber
band, serves to keep the lower end of the apparatus in position.
The period of vibration of the pendulum bar is adjustable to any
desired magnitude by shifting in or out the attached weights, or by
tightening or relaxing the cords which pull the upper ends of the
spiral springs.

blackened behind, I see coloured rings ; the phenomenon will be shown to you on the screen by means of the electric light reflected from the space of air between the two pieces of glass. This phenomenon was first observed by Sir Isaac Newton, and was first explained by the undulatory theory of light. [Newton's rings are now shown on the screen before you by reflected electric light.]　If I press the glasses together, you see a dark spot in the centre ; the rings appear round it, and there is a dark centre with irregularities.　Pressure is required to produce that spot.　Why ?　The answer generally given is, because glass repels glass at a distance of two or three wave-lengths of light ; say at a distance of 1/5000th of a centimetre.　I do not believe that for a moment.　The seeming repulsion comes from shreds or particles of dust between them. The black spot in the centre is a place where the distance between them is less than a quarter of a wave-length.　Now the wave-length for yellow light is about 1/17,000th of a centimetre, and the quarter of 1/17,000th is about 1/70,000th.

The place where you see the middle of that black circle is an air-space, with the distance apart less than 1/70,000th of a centimetre. Passing from this black spot to the first ring of maximum light, add half a wave-length to the distance, and we can tell what is the distance between the two pieces of glass at this place ; add another half wave-length, and we come to the next maximum of light again ; but the colour prevents us speaking very definitely because we have a number of different wave-lengths concerned. I will simplify that by reducing it all to one colour, red, by interposing a red glass. You have now one colour, but much less light altogether, because this glass only lets through homogeneous red light, or not much besides. Now look at what you see on the screen, and you have unmistakeable evidence of fulcrums of dust between the glass surfaces. When I put on the screw, I whiten the central black spot by causing the elastic glass to pivot, as it were, round the innumerable little fulcrums constituted by the molecules of dust ; and the pieces of glass are pressed

not against one another, but against these fulcrums. There are innumerable—say thousands—of
little particles of dust jammed between the glass
surfaces, some of them of perhaps 1/3000th of a
centimetre in diameter, say 5 or 6 wave-lengths.
If you lay one piece of glass on another, you
think you are pressing glass on glass, but it is
nothing of the kind : it is glass on dust. This
is a very beautiful phenomenon, and my object
in showing this experiment was simply because
it gives us a linear measure bringing us down
at once to 1/100,000th of a centimetre.

Now I am going to enter a little into detail
regarding the reasons that four lines of argument
give us for assigning a limit to the smallness of
the molecules of matter. I shall take contact
electricity first, and very briefly.

If I take these two pieces of zinc and copper
and touch them together at the two corners, they
become electrified, and attract one another with
a perfectly definite force, of which the magnitude
is ascertained from absolute measurements in connection with the well-established doctrine of contact

M

electricity. I do not feel it, because the force is very small, but you may do the thing in a measured way ; you may place a little metallic knob or projection of 1/100,000th of a centimetre, on one of them, and lean the other against it. Let there be three such little metal feet put on the copper ; now touch the zinc plate with one of them, and turn it gradually down till it comes to touch the other two. In this position, with an air-space of 1/100,000th of a centimetre between them, there will be positive and negative electricity on the zinc and copper surfaces respectively, of such quantities as to cause a mutual attraction amounting to 2 grammes weight per square centimetre. The amount of work done by the electric attraction upon the plates while they are being allowed to approach one another with metallic connection between them at the corner first touched, till they come to the distance of 1/100,000th of a centimetre, is 2/100,000ths of a centimetre-gramme, supposing the area of each plate to be one square centimetre.

Let me read you this statement taken from an

article which was published thirteen years ago in *Nature.*[1]

"Now let a second plate of zinc be brought by a similar process to the other side of the plate of copper; a second plate of copper to the remote side of this second plate of zinc, and so on till a pile is formed consisting of 50,001 plates of zinc and 50,000 plates of copper, separated by 100,000 spaces, each plate and each space 1/100,000th of a centimetre thick. The whole work done by electric attraction in the formation of this pile is two centimetre-grammes.

"The whole mass of metal is eight grammes. Hence the amount of work is a quarter of a centimetre-gramme per gramme of metal. Now, 4030 centimetre-grammes of work, according to Joule's dynamical equivalent of heat, is the amount required to warm a gramme of zinc or copper by one degree Centigrade. Hence the work done by the electric attraction could warm the substance by only 1/16,120th of a degree. But now let the

[1] See article "On the Size of Atoms," published in *Nature*, vol. i. p. 551; printed in Thomson and Tait's *Natural Philosophy*, Second Edition, 1883, vol. i. part II, Appendix F.

M 2

thickness of each piece of metal and of each intervening space be 1/100,000,000th of a centimetre, instead of 1/100,000th. The work would be increased a million-fold unless 1/100,000,000th of a centimetre approaches the smallness of a molecule. The heat equivalent would therefore be enough to raise the temperature of the material by 62° This is barely, if at all, admissible, according to our present knowledge, or, rather, want of knowledge, regarding the heat of combination of zinc and copper. But suppose the metal plates and intervening spaces to be made yet four times thinner, that is to say, the thickness of each to be 1/400,000,000th of a centimetre. The work and its heat equivalent will be increased sixteen-fold. It would therefore be 990 times as much as that required to warm the mass by one degree Centigrade, which is very much more than can possibly be produced by zinc and copper in entering into molecular combination. Were there in reality anything like so much heat of combination as this, a mixture of zinc and copper powders would, if melted in any one spot, run together, generating more than enough heat to melt each throughout ; just as a large quantity of gunpowder if ignited in any one spot burns throughout without fresh application of heat. Hence plates of zinc and copper of 1/300,000,000th of a centimetre thick, placed close together alternately, form a near approximation to a chemical combination, if indeed such thin plates could be made without splitting atoms."

In making brass, if we mix zinc and copper together we find no very manifest signs of chemical affinity at all ; there is not a great deal of heat developed ; the mixture does not become warm, *it does not explode.* Hence we can infer certainly that contact-electricity action ceases, or does not go on increasing according to the same law, when the metals are subdivided to something like 1/100,000,000th of a centimetre. Now this is an exceedingly important argument. I have more decided data as to the actual magnitude of atoms or molecules to bring before you presently, but I have nothing more decided in *giving for certain a limit to supposable smallness.* We cannot reduce zinc and copper beyond a certain thickness without putting them into a condition in which they lose their properties as distinct solid metals, and in which, if put together, we should *not* find the same attraction as we should calculate upon from the thicker plates. I think it is impossible, consistently with the knowledge we have of chemical affinities and of the effect of melting zinc and copper together, to admit that a piece of copper or zinc could be divided to a thinness of much less, if at all less, than 1/100,000,000th of a centimetre without separating the atoms or dividing the molecules, or doing away with the composition which constitutes as a whole the solid metal. In short, the constituents as it were of bricks, or molecules, or atoms, of which copper and zinc are built up, cannot be much, if at all, less than 1/100,000,000th of a centimetre in diameter, and may be considerably greater.

Similar conclusions result from that curious and most interesting phenomenon, the soap-bubble. Philosophers, old and young, who occupy themselves with soap-bubbles, have one of the most interesting subjects of physical science to admire. Blow a soap-bubble and look at it,— you may study all your life perhaps, and still learn lessons in physical science from it. You will now see on the screen the image of a soap-film in a ring of metal. The light is reflected from the film filling that ring, and made to focus on the screen. It will show, as you see, colours analogous to those of Newton's rings. As you see it the image is upside down. The liquid streams down (up in the image), and thins away from the highest point of the film. First we see that brilliant green colour. It will become thinner and thinner there, and will pass through beautiful gradations of colour till you see, as now, a deep red, then much lighter, till it becomes a dusky, yellowish-white, then green, and blue, and deep violet, and lastly black, but after you see the black spot it very soon bursts. The film itself seems to begin to lose its tension when its thickness gets considerably less than a quarter of the wave-length of yellow light, which is the thickness for the dusky white, preceding the

final black. When you are washing your hands,
you may make and deliberately observe a film
like this, in a ring formed by the forefingers and
thumbs of two hands, and watch the colours.
Whenever you begin to see a black spot or
several black spots, the film soon after breaks.
The film retains its strength until we come to
the black spot, where the thickness is clearly
much less than 1-60,000th of a centimetre, which
is the thickness of the dusky white.[1]

Newton, in the following passage in his ' Optics '
(pp. 187 and 191 of edition 1721, Second Book,

---

[1] Since this lecture was delivered a paper "On the Limiting
Thickness of Liquid Films," by Professors Reinold and Rücker,
has been communicated to the Royal Society, and an abstract has
been published in the *Proceedings*, April 19, 1883. The authors
give the following results for the thickness of a black film of the
liquids specified :—

Liquid.	Method.	Mean Thickness.
Plateau's " Liquide	Electrical.	$\cdot119 \times 10^{-5}$ cm.
Glycérique."	Optical.	$\cdot107$,,
Soap Solution.	Electrical.	$\cdot117$,,
	Optical.	$\cdot121$,,

The thickness, therefore, of a film of the "liquide glycérique"
and that of a film of a soap solution containing no glycerine are
nearly the same, and about 1/50th of the wave-length of sodium
light.

Part I.) tells more of this important phenomenon of the black spot than is known to many of the best of modern observers.

"Obs. 17.—If a bubble be blown with water, first made tenacious by dissolving a little soap in it, it is a common observation that after a while it will appear tinged with a variety of colours. To defend these bubbles from being agitated by the external air (whereby their colours are irregularly moved one among another so that no accurate observation can be made of them), as soon as I had blown any of them I covered it with a clear glass, and by that means its colours emerged in a very regular order, like so many concentric rings encompassing the top of the bubble. And as the bubble grew thinner by the continual subsiding of the water, these rings dilated slowly and overspread the whole bubble, descending in order to the bottom of it, where they vanished successively. In the meanwhile, after all the colours were emerged at the top, there grew in the centre of the rings a small round black spot like that in the first observation, which continually dilated itself till it became sometimes more than one-half or three-quarters of an inch in breadth before the bubble broke. At first I thought there had been no light reflected from the water in that place, but observing it more curiously I saw within it several smaller round spots, which appeared much blacker and darker than the rest, whereby I knew that there was some reflection at the other places which were not so dark as those spots. And by further trial I found that I could see the images of some things (as of a candle or the sun) very faintly reflected, not only from the great black spot, but also from the little darker spots which were within it.

" Obs. 18.—If the water was not very tenacious, the black spots would break forth in the white without any sensible intervention of the blue. And sometimes they would break forth within the precedent ·yellow, or red, or perhaps within the blue of the second order, before the intermediate colours had time to display themselves."

Now I have a reason, an irrefragable reason, for

saying that the film cannot keep up its tensile strength to 1/100,000,000th of a centimetre, and that is, that the work which would be required to stretch the film a little more than that would be enough to drive it into vapour.

The theory of capillary attraction shows that when a bubble—a soap-bubble, for instance—is blown larger and larger, work is done by the stretching of the film which resists extension as if it were an elastic membrane with a constant contractile force. This contractile force is to be reckoned as a certain number of units of force per unit of breadth. Observation of the ascent of water in capillary tubes shows that the contractile force of a thin film of water is about 16 milligrammes weight per millimetre of breadth. Hence the work done in stretching a water film to any degree of thinness, reckoned in millimetre-milligrammes, is equal to sixteen times the number of square millimetres by which the area is augmented, provided the film is not made so thin that there is any sensible diminution of its contractile force. In an article " On the Thermal

Effect of Drawing out a Film of Liquid," published in the *Proceedings* of the Royal Society for April 1858, [*Math. and Phys. Papers,* vol. iii. Art. XCV.], I have proved from the second law of thermodynamics that about half as much more energy, in the shape of heat, must be given to the film, to prevent it from sinking in temperature while it is being drawn out. Hence the intrinsic energy of a mass of water in the shape of a film kept at constant temperature increases by 24 millimetre-milligrammes for every square millimetre added to its area.

Suppose, then, a film to be given with the thickness of a millimetre, and suppose its area to be augmented ten thousand-and-one fold : the work done per square millimetre of the original film, that is to say, per milligramme of the mass, would be 240,000 millimetre-milligrammes. The heat equivalent to this is more than half a degree Centigrade ($0°57°$) of elevation of temperature of the substance. The thickness to which the film is reduced on this supposition is very approximately $1/10,000$th of a millimetre. The commonest observation on the soap-bubble shows that there is no sensible diminution of contractile force by reduction of the thickness to 1/10,000th of a millimetre; inasmuch as the thickness which gives the first maximum brightness, round the black spot seen where the film is thinnest, is only about 1/6000th of a millimetre.

The very moderate amount of work shown in the preceding estimates is quite consistent with this reduction. But suppose now the film to be further stretched until its thickness is reduced to 1/10,000,000th of a millimetre (1/100,000,000th of a centimetre). The work spent in doing this is one thousand times more than that which we have just calculated. The heat equivalent is 570 times the quantity required to raise the temperature of the liquid by 1° Centigrade. This is far more than we can admit as a possible amount of work done in the extension of a liquid film. It is more than the amount of work which, if spent on the liquid, would convert it into vapour at ordinary atmospheric pressure. The conclusion is unavoidable, that a water-film falls off greatly in its contractile force before it is reduced to a thickness of 1/10,000,000th of a millimetre. It is scarcely possible, upon any conceivable molecular theory, that there can be any considerable falling off in the contractile force as long as there are several molecules in the thickness. It is therefore probable that there are not several molecules in a thickness of 1/10,000,000th of a millimetre of water.

Now when we are considering the subdivision of matter, look at those beautiful colours which you see in this little casket, left, I believe, by Professor Brande to the Royal Institution. It contains polished steel bars, coloured by having been raised to different degrees of heat, as in the process of annealing hard-tempered steel. These colours, produced by heat on other polished metals besides steel, are due to thin films of transparent oxide, and their tints, as those of the soap-bubble and of the thin space of air in " Newton's rings," depend on the thickness of the film, which, in the case of oxidisable metals, forms, by combination with the oxygen of the air under the influence of heat, a true surface-burning.

You are all familiar with the brilliant and beautifully distributed fringes of heat-colours on polished steel grates and fire-irons escaping that unhappy rule of domestic æsthetics which too often keeps those articles glittering and cold and useless, instead of letting them show the exquisite play of warm colouring naturally and inevitably brought out when they are used in the work which is their reason for existence. The thickness of the film of oxide which gives the first perceptible colour, a very pale orange or buff tint, due to the enfeeblement or extinction of violet light and enfeeblement of blue and less enfeeblement of the other colours in order, by interference of the reflections from the two surfaces of the film, is about $1/100,000$th of a centimetre, being something less than a quarter wavelength of violet light.

The exceedingly searching and detective efficacy of electricity comes to our aid here, and by the force, as it were, spread through such a film, proves to us the existence of the film when it is considerably thinner than that $1/100,000$th of a centimetre, when in fact it is so very thin as to produce absolutely no perceptible effect on the reflected light, that is to say, so thin as to be absolutely invisible. If, in the apparatus for measuring contact electricity, of which the drawing is before you (*Nature*, vol. xxiii. p. 567), two plates of freshly polished copper be placed in the Volta condenser, a very perfect zero of effect is obtained. If, then, one of the plates be taken out, heated slightly by laying it on a piece of hot iron, and then allowed to cool again and replaced in the Volta condenser, it is found that negative electricity becomes condensed on the surface thus treated, and positive electricity on the bright copper surface facing it, when the two are in metallic connection. If the same process be repeated with somewhat higher temperatures, or somewhat longer times of exposure to it, the electrical difference is augmented. These effects are very sensible before any perceptible tint appears on the copper surface as modified by heat. The effect goes on increasing with higher and higher temperatures of the heating influence, until oxide tints begin to appear, commencing with buff, and going on through a ruddier colour to a dark-blue slate colour, when no further heating seems to augment the effect. The greatest contact-electricity effect which I thus obtained between a bright freshly polished copper surface and an opposing face of copper, rendered almost black by oxidation, was such as to require for the neutralising potential in my mode of experimenting [1] about one-half of the potential of a Daniell's cell.

Some not hitherto published experiments with polished silver plates, which I made fifteen years ago, showed me very startlingly an electric influence from a quite infinitesimal whiff of iodine vapour. The effect on the contact-electricity quality of the surface seems to go on continuously from the first lodgment, to all other tests quite

[1] First described in a letter to Joule, published in the *Proceedings of the Literary and Philosophical Society of Manchester*, of Jan. 21, 1862, where also I first pointed out the demonstration of a limit to the size of molecules from measurements of contact-electricity. The mode of measurement is more fully described in the article of *Nature* (vol. xxiii. p. 567) referred to above.

N

imperceptible, of a few atoms or molecules of the attacking substance (oxygen, or iodine, or sulphur, or chlorine, for example), and to go on increasing until some such thickness as 1/30,000th or 1/40,000th of a centimetre is reached by the film of oxide or iodide, or whatever it may be that is formed.

The subject is one that deserves much more of careful experimental work and measurement than has hitherto been devoted to it. I allude to it at present to point out to you how it is that by this electric action we are enabled as it were to sound the depth of the ocean of molecules attracted to the metallic surface by the vapour or gas entering into combination with it.

When we come to thicknesses of considerably less than a wave-length we find solid metals becoming transparent. Through the kindness of Prof. Dewar I am able to show you some exceedingly thin films of measured thicknesses of platinum, gold, and silver, placed on glass plates. The platinum is of $1 \cdot 9 \times 10^{-5}$ cms. thickness, and is quite opaque; but here is a gold film of about the same thickness, which is transparent to the electric light, as you see, and transmits the beautiful green colour which you see on the screen. The thickness of this gold (1·9, or nearly 2×10^{-5} cm.) is just half the wave-length of violet light in air. This transparent gold, transmitting green light to the screen as you see, at the same time reflects yellow light to the ceiling. Now I will show you the silver. It is thinner, being only $1·5 \times 10^{-5}$ of a centimetre thick, or ⅜ths of the air-wave-length of violet light. It is quite opaque to the electric light so far as our eyes allow us to judge, and reflects all the light up to the ceiling. It is not wonderful that it should be opaque; we might wonder if it were otherwise; but there is an invisible ultra-violet light of a small range of wave-lengths, including a zinc line of air-wave-length $3·4 \times 10^{-5}$ cms., which this silver film transmits. For that particular light the silver film of $1·5 \times 10^{-5}$ cms. thickness is transparent. The image which you now see on the screen is a magic-lantern representation of the self-photographed spectrum of light that actually came through that silver. You

N 2

see the zinc line very clear across it near its middle. Here then we have gold and silver transparent. The silver is opaque for all except that very definite light of wave-lengths from about 3·07 to 3·32 10^{-5} cms.

The different refrangibility of different colours is a result of observation of vital importance in the question of the size of atoms. You now see on the screen before you a prismatic spectrum, a well-known phenomenon produced by the differences of the refractions of the different colours in traversing the prism. The explanation of it in the undulatory theory of light has taxed the powers of mathematicians to the utmost. Look first, however, to what is easy and is made clear by that diagram (Fig. 35) before you, and you will easily understand that refraction depends on difference of velocity of propagation of light in the two transparent mediums concerned. The angles in the diagram are approximately correct for refraction at an interface between air or vacuum and flint glass; and you see that in this case the velocity of propagation is less in the

denser medium. The more refractive medium
(not always the denser) of the two has the less
velocity for light transmitted through it. The
"refractive index" of any transparent medium is
the ratio of the velocity of propagation in the

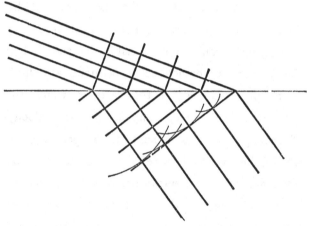

FIG. 35.—Diagram of Huyghen's construction for wave front of refracted light;
drawn for light passing from air to flint glass.

ether to the velocity of propagation in the transparent substance.

Now that the velocity of the propagation of
light should be different in different mediums,
and should in most cases be smaller in the denser
than in the less dense medium, is quite what we

should, according to dynamical principles, expect from any conceivable constitution of the luminiferous ether and of palpable transparent substance. But that the velocity of propagation in any one transparent substance should be different for light of different colours, that is to say, of different periods of vibration, is not what we should expect, and could not possibly be the fact if the medium is homogeneous, without any limit as to the smallness of the parts of which the qualities are compared. The fact that the velocity of propagation *does* depend on the period, gives what I believe to be irrefragable proof that the substance of palpable transparent matter, such as water, or glass, or the bisulphide of carbon of this prism, whose spectrum is before you, is not infinitely homogeneous ; but that, on the contrary, if contiguous portions of any such medium, any medium in fact which can give the prismatic colours, be examined at intervals not incomparably small in comparison with the wavelengths, utterly heterogeneous quality will be discovered ; such heterogeneousness as that which we understand, in palpable matter, as the difference between solid and fluid, or between substances differing enormously in density; or such heterogeneousness as differences in velocity and in direction of motion, in different positions of a vortex ring in an homogeneous liquid; or such differences of material occupying the space examined, as we find in a great mass of brick building when we pass from brick to brick through mortar (or through *void*, as we too often find in Scotch-built domestic brick chimneys).

Cauchy was, I believe, the first of mathematicians or naturalists to allow himself to be driven to the conclusion that the refractive dispersion of light can only be accounted for by a finite degree of molecular coarse-grainedness in the structure of the transparent refracting matter; and as, however we view the question, and however much we may feel compelled to differ from the details of molecular structure and molecular inter-action assumed by Cauchy, we remain more and more surely fortified in his conclusion, that finite coarse-grainedness of transparent palpable matter is the cause of the difference in the velocities of different colours of light propagated through it, we must regard Cauchy as the discoverer of the dynamical theory of the prismatic colours.

But now we come to the grand difficulty of Cauchy's theory.[1] Look at this little table (Table II.), and you will see in the heading the

TABLE II.—VELOCITY (V) ACCORDING TO NUMBER (N) OF PARTICLES IN WAVE-LENGTH.

$N.$	$V\left(= 100\ \dfrac{\sin (\pi/N)}{\pi/N}\right).$
2	63·64
4	90·03
8	97·45
12	98·86
16	99·36
20	99·59
∞	100·00

formula which gives the velocity, in terms of the number of particles to the wave-length, supposing the medium to consist of equal particles arranged

[1] For an account of the dynamical theory of the "Dispersion of Light," see *View of the Undulatory Theory as applied to the Dispersion of Light*, by the Rev. Baden Powell, M.A., &c. (London, 1841.)

in cubic order, and each particle to attract its six nearest neighbours, with a force varying directly as the excess of the distance between them, above a certain constant line (the length of which is to be chosen, according to the degree of compressibility possessed by the elastic solid, which we desire to represent by a crowd of mutually interacting molecules). If you suppose particles of real matter arranged in the cubic order, and six steel wire spiral springs, or elastic india-rubber bands, to be hooked on to each particle and stretched between it and its six nearest neighbours, the postulated force may be produced in a model with all needful accuracy ; and if we could but successfully *wish* the theatre of the Royal Institution conveyed to the centre of the earth and kept there for five minutes, I should have great pleasure in showing you a model of an elastic solid thus constituted, and showing you waves propagated through it, as are waves of light in the luminiferous ether. Gravity is the inconvenient accident of our actual position which prevents my showing it to you here just now.

But instead, you have these two wave-models (see Fig. 34, page 157), each of which shows you the displacement and motion of a line of particles in the propagation of a wave through our imaginary three-dimensional solid; the line of molecules chosen being those which in equilibrium are in one direct straight line of the cubic arrangement, and the supposed wave having its wave front perpendicular to this line, and the direction of its vibration the direction of one of the other two direct lines of the cubic arrangement.

You have also before you this series of diagrams (Figs. 36 to 41) of waves in a molecularly-constituted elastic solid. These two diagrams (Figs. 36 and 37) illustrate a wave in which there are twelve molecules in the wave-length; this one (Fig. 36) showing (by the length and position of the arrows) the magnitude and direction of velocity of each molecule at the instant when one of the molecules is on the crest of the wave, or has reached its maximum displacement; that one (Fig. 37) showing the magnitude and direction of the velocities after the wave has advanced such

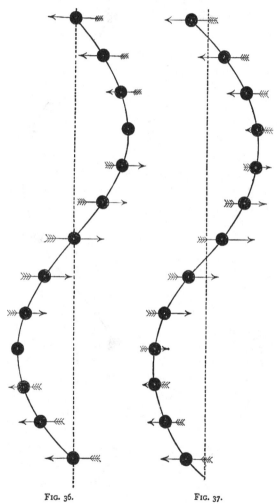

FIG. 36. FIG. 37.

Twelve particles in Wave-Length.

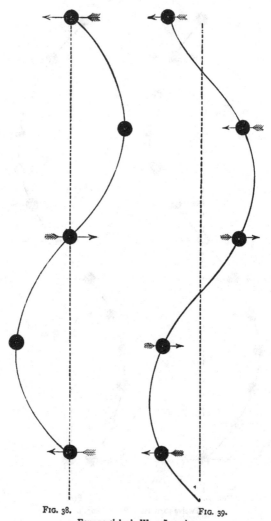

Fig. 38.   Fig. 39.

Four particles in Wave-Length.

a distance as (in this case equal to 1/24th of the wave-length) to bring the crest of the wave to midway between two molecules. This pair of diagrams (Figs. 38 and 39) shows the same for a wave having four molecules in the wave-length, and this pair (Figs. 40 and 41) for a wave having two molecules in the wave-length.

The more nearly this critical case is approached, that is to say, the shorter the wave-length down to the limit of twice the distance from molecule to molecule, the less becomes the difference between the two configurations of motion constituted by waves travelling in opposite directions. In the extreme or critical case the difference is annulled, and the motion is not a wave motion, but a case of what is often called "standing vibration." Before I conclude this evening I hope to explain in detail the kind of motion which we find instead of wave-motion (become mathematically imaginary), when the vibrational period of the exciter is anything less than the critical value, because this case is of extreme importance and interest in physical optics, according to

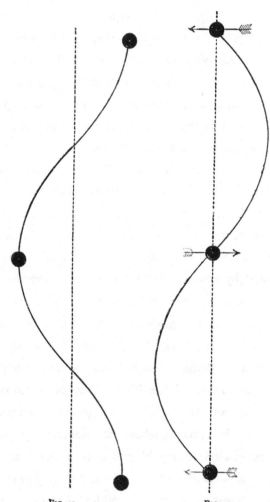

FIG. 40.
Two particles in Wave-Length.
FIG. 41.

Stokes's hitherto unpublished explanation of phosphorescence

This supposition of each molecule acting with direct force only on its nearest neighbours is not exactly the postulate on which Cauchy works. He supposes each molecule to act on all around it, according to some law of rapid decrease as the distance increases; but this must make the influence of coarse-grainedness on the velocity of propagation smaller than it is on the simple assumption realised in the models and diagrams before you, which therefore represents the extreme limit of the efficacy of Cauchy's unmodified theory to explain dispersion.

Now, by looking at the little table (Table II. p. 184) of calculated results, you will see that, with as few as 20 molecules in the wave-length, the velocity of propagation is $99\frac{1}{2}$ per cent. of what it would be with an infinite number of molecules; hence the extreme difference of propagational velocity, accountable for by Cauchy's unmodified theory in its idealised extreme of mutual action limited to nearest neighbours, amounts to 1/200th.

Now look at this table (Table III.) of refractive indices, and you see that the difference of velocity

TABLE III.—TABLE OF REFRACTIVE INDICES.

Line of Spectrum.	Material.				
	Hard Crown Glass.	Extra dense Flint Glass.	Water at 15° C.	Carbon Disulphide at 11° C.	Alcohol at 15° C.
A	1·5118	1·6391	1·3284	1·6142	1·3600
B	1·5136	1·6429	1·3300	1·6207	1·3612
C	1·5146	1·6449	1·3307	1·6240	1·3621
D	1·5171	1·6504	1·3324	1·6333	1·3638
E	1·5203	1·6576	1·3347	1·6465	1·3661
b	1·5210	1·6591	...	...	...
F	1·5231	1·6442	1·3366	1·6584	1·3683
G	1·5283	1·6770	1·3402	1·6836	1·3720
h	1·5310	1·6836	...	...	...
H	1·5328	1·6886	1·3431	1·7090	1·3751

The numbers in the first two columns were determined by Dr. Hopkinson, those in the last three by Messrs. Gladstone and Dale. The index of refraction of air for light near the line E is 1·000294.

of red light A, and of violet light H, amounts in carbon disulphide to 1/17th; in dense flint glass to nearly 1/30th; in hard crown glass to 1/73rd; and in water and alcohol to rather more than 1/100th. Hence, none of these substances can have so many as 20 molecules in the wave-length, if dispersion is to be accounted for

by Cauchy's unmodified theory, and by looking
back to the little table of calculated results
(Table II. p. 184), you will see that there could not
be more than 12 molecules in the wave-length
of violet light in water or alcohol; say 10 in
hard crown glass; 8 in flint glass; and in
carbon disulphide actually not more than 4
molecules in the wave-length, if we are to depend
upon Cauchy's unmodified theory for the explanation of dispersion. So large coarse-grainedness
of ordinary transparent bodies, solid or fluid, is
quite untenable. Before I conclude, I intend to
show you, from the kinetic theory of gases, a
superior limit to the size of molecules, according
to which, in glass or in water, there is probably
something like 600 molecules to the wave-length,
and almost certainly *not fewer* than 200, or 300, or
400. But even without any such definite estimate
of a superior limit to the size of molecules, there
are many reasons against the admission that it is
probable or possible there can be only four, or
five, or six, to the wave-length. The very drawing,
by Nobert, of 4000 lines on a breadth of a milli-

O

metre, or at the rate of 40,000 to the centimetre, or about two to the ether wave-length of blue (F) light,[1] seems quite to negative the idea of any such possibility of only five or six molecules to the wave-length, even if we were not to declare against it from theory and observation of the reflection of light from polished surfaces.

We must then find another explanation of dispersion ; and I believe there is another explanation I believe that, while giving up Cauchy's unmodified theory of dispersion, we shall find that the same general principle is applicable, and that by imagining each molecule to be loaded in a certain definite way by elastic connection with heavier matter—each molecule of the ether to have, in palpable transparent matter, a small fringe so to speak of particles, larger and larger in their successive order, elastically connected with it—we shall have a rude mechanical explanation, realisable by the notably easy addition of the

[1] Loschmidt, quoting from the Zollvereins department of the London International Exhibition of 1862, p. 83, and from Harting "On the Microscope," p. 881, *Sitzungsberichte der Wiener Akademie Math. Phys.*, 1865, vol. iii.

proper appliances to the dynamical models before you to account for refractive dispersion in an infinitely fine-grained structure. It is not seventeen hours since I saw the possibility of this explanation. I think I now see it perfectly, but you will excuse my not going into the theory more fully under the circumstances.[1] The difficulty of Cauchy's theory has weighed heavily upon me when thinking of bringing this subject before you. I could not bring it before you and say there are only four particles in the wave-length, and I could not bring it before you without saying there is some other explanation. I believe another explanation is distinctly to be had in the manner I have slightly indicated.

Now look at those beautiful distributions of colour on the screen before you. They are diffraction spectrums from a piece of glass ruled with 2000 lines to the inch. And again look, and

[1] Further examination has seemed to me to confirm this first impression ; and in a paper " On the Dynamical Theory of Dispersion," read before the Royal Society of Edinburgh on the 5th of March, I have given a mathematical investigation of the subject.—W. T., March 16, 1883.

you see one diffraction spectrum by reflection from one of Rutherford's gratings, in which there are 17,000 lines to the inch on polished speculum-metal. The explanation by "interference" is substantially the same as that which the undulatory theory gives for Newton's rings of light reflected from the two surfaces, which you have already seen. Where light-waves from the apertures between the successive bars of the grating reach the screen in the same phase, they produce light; there, again, where they are in opposite phases, they produce darkness.

The beautiful colours which are produced depend on the places of conspiring and opposing vibrations on the screen, being different for light-waves of different wave-lengths ; and it was by the measurement of the dimensions of a diffraction spectrum such as the first set you saw (or of finer spectrums from coarser gratings) that Fraunhofer first determined the wave-lengths of the different colours.

I have now, closely bearing on the question of the size of atoms, thanks to Dr. Tyndall, a most

beautiful and interesting experiment to show you
—the artificial "blue sky," produced by a very
wonderful effect of light upon matter, which he
discovered. We have here an empty glass tube
—it is "optically void." A beam of electric light
passes through it now, and you see nothing. Now
the light is stopped, and we admit vapour of
carbon disulphide into the tube. There is now
introduced some of this vapour to about 3 inches
pressure, and there is also introduced, to the
amount of 15 inches pressure, air impregnated
with a little nitric acid, making in all rather less
than the atmospheric pressure. What is to be
illustrated here is the presence of molecules of
substances produced by the decomposition of
carbon disulphide by the light. At present you
see nothing in the tube ; it still continues to be,
as before the admission of the vapours, optically
transparent : but gradually you will see an exquisite blue cloud. That is Tyndall's "blue sky."
You see it now. I take a Nicol's prism, and by
looking through it I find the azure light coming
from the vapours in any direction perpendicular

to the exciting beam of light to be very completely polarised in the plane through my eye and the exciting beam. It consists of light-vibrations in one definite direction, and that, as finally demonstrated by Professor Stokes, it seems to me beyond all doubt, through reasoning on this phenomenon of polarisation,[1] which he had observed in various

---

[1] Extract from Professor Stokes's paper "On the Change of Refrangibility of Light," read before the Royal Society, May 27th, 1852, and published in the *Transactions* for that date :—

"§ 183. Now this result appears to me to have no remote bearing on the question of the directions of the vibration in polarised light. So long as the suspended particles are large compared with the waves of light, reflection takes place as it would from a portion of the surface of a large solid immersed in the fluid, and no conclusion can be drawn either way. But if the diameters of the particles be small compared with the length of a wave of light, it seems plain that the vibrations in a reflected ray cannot be perpendicular to the vibrations in the incident ray. Let us suppose for the present, that in the case of the beams actually observed, the suspended particles were small compared with the length of a wave of light. Observation showed that the reflected ray was polarised. Now all the appearances presented by a plane polarised ray are symmetrical with respect to the plane of polarisation. Hence we have two directions to choose between for the direction of the vibrations in the reflected ray, namely, that of the incident ray, and a direction perpendicular to both the incident and the reflected rays. The former would be necessarily perpendicular to the directions of vibration in the incident ray, and therefore we are obliged to choose the latter, and consequently to suppose that the vibrations of

experimental arrangements giving minute solid or liquid particles scattered through a transparent medium, must be the direction perpendicular to the plane of polarisation.

What you are now about to see, and what I tell you I have seen through the Nicol's prism, is due to what I may call secondary or derived waves of light diverging from very minute liquid spherules, condensed in consequence of the chemical decomposing influence exerted by the beam of light on the matter in the tube, which was all gaseous when the light was first admitted.

plane polarised light are perpendicular to the plane of polarisation, since experiment shows that the plane of polarisation of the reflected ray is the plane of reflection. According to this theory, if we resolve the vibrations in the [horizontal] incident ray horizontally and vertically, the resolved parts will correspond to the two rays, polarised respectively in and perpendicularly to the plane of reflection, into which the incident ray may be conceived to be divided, and of these the former alone is capable of furnishing a . . . . . ray reflected vertically upwards [to be seen by an eye above the line of the incident ray, and looking vertically downwards]. And, in fact, observation shows that, in order to quench the dispersed beam, it is sufficient, instead of analysing the reflected light, to polarise the incident light in a plane perpendicular to the plane of reflection."

To understand these derived waves, first you must regard them as due to motion of the ether round each spherule ; the spherule being almost absolutely fixed, because its density is enormously greater than that of the ether surrounding it. The motion that the ether had in virtue of the exciting beam of light alone, before the spherules came into existence, may be regarded as being compounded with the motion of the ether relatively to each spherule, to produce the whole resultant motion experienced by the ether when the beam of light passes along the tube, and azure light is seen proceeding from it laterally. Now this second component motion is clearly the same as the whole motion of the ether would be, if the exciting light were annulled and each spherule kept vibrating in the opposite direction, to and fro through the same range as that which the ether in its place had, in virtue of the exciting light, when the spherule was not there.

Supposing now, for a moment, that without any exciting beam at all, a large number of minute spherules are all kept vibrating through very small ranges[1] parallel to one line. If you place your eye in the plane through the length of the tube and perpendicular to that line, you will see light from all parts of the tube, and this light which you see will consist of vibrations

[1] In the following question of the recent Smith's Prize Examination at Cambridge (paper of Tuesday, Jan. 30, 1883), the dynamics of the subject, and particularly the motion of the ether produced by keeping a single spherule embedded in it vibrating to and fro in a straight line, are illustrated in parts (a) and (d) :—

"8. (a) From the known phenomenon that the light of a cloudless blue sky, viewed in any direction perpendicular to the sun's direction, is almost wholly polarised in the plane through the sun, assuming that this light is due to particles of matter of diameters small in comparison with the wave-length of light, prove that the direction of the vibrations of plane polarised light is perpendicular to the plane of polarisation.

"(b) Show that the equations of motion of a homogeneous isotropic elastic solid of unit density, are

$$\frac{d^2a}{dt^2} = (k + \tfrac{1}{3} n) \frac{d\delta}{dx} + n\nabla^2 a,$$

$$\frac{d^2\beta}{dt^2} = (k + \tfrac{1}{3} n) \frac{d\delta}{dy} + n\nabla^2\beta,$$

$$\frac{d^2\gamma}{dt^2} = (k + \tfrac{1}{3} n)\frac{d\delta}{dz} + n\nabla^2\gamma,$$

where k denotes the modulus of resistance to compression ; n the rigidity-modulus ; a, β, γ, the components of displacement at (x, y, z, t); and

$$\delta = \frac{da}{dx} + \frac{d\beta}{dy} + \frac{d\gamma}{dz},$$

$$\nabla^2 = \frac{d^2}{dx^2} + \frac{d^2}{dy^2} + \frac{d^2}{dz^2}.$$

parallel to that line. But if you place your eye *in* the line of the vibration of a spherule, situated about the middle of the tube, you will see no light in that direction; but keeping your eye in the same position, if you look obliquely towards either end of the tube, you will see light fading into darkness, as you turn your eye from either end towards the middle. Hence, if the exciting

" (*c*) Show that every possible solution is included in the following :—

$$a = \frac{d\phi}{dx} + u, \quad \beta = \frac{d\phi}{dy} + v, \quad \gamma = \frac{d\phi}{dz} + w,$$

where *u, v, w,* are such that

$$\frac{du}{dx} + \frac{dv}{dy} + \frac{dw}{dz} = 0.$$

" Find differential equations for the determination of ϕ, *u, v, w.* Find the respective wave-velocities for the ϕ-solution, and for the (*u, v, w*)-solution.

" (*d*) Prove the following to be solutions, and interpret each for values of $r\,[\sqrt{\ }\,(x^2 + y^2 + z^2)]$ very great in comparison with λ (the wave-length).

(1) $\begin{cases} a = \dfrac{d\phi}{dx}, \quad \beta = \dfrac{d\phi}{dy}, \quad \gamma = \dfrac{d\phi}{dz}, \\[2mm] \text{where } \phi = \dfrac{1}{r}\sin\dfrac{2\pi}{\lambda}\,[r - t\sqrt{\ }\,(k + \tfrac{4}{3}\,n)]. \end{cases}$

(2) $\begin{cases} a = 0, \quad \beta = -\dfrac{d\psi}{dz}, \quad \gamma = \dfrac{d\psi}{dy} \\[2mm] \text{where } \psi = \dfrac{1}{r}\sin\dfrac{2\pi}{\lambda}\,[r - t\sqrt{\ }\,n]. \end{cases}$

(3) $a = \left(\dfrac{2\pi}{\lambda}\right)\psi + \dfrac{d^2\psi}{dx^2}, \quad \beta = \dfrac{d^2\psi}{dxdy}, \quad \gamma = \dfrac{d^2\psi}{dxdz}.$

beam be of plane polarised light—that is to say, light of which all the vibrations are parallel to one line—and if you look at the tube in the direction perpendicular to this line and to the length of the tube, you will see light of which the vibrations will be parallel to that same line. But if you look at the tube in any direction parallel to this line, you will see no light ; and the line along which you see no light is the direction of the vibrations in the exciting beam ; and this direction, as we now see, is the direction perpendicular to what is technically called the plane of polarisation of the light. Here, then, you have Stokes' *experimentum crucis* by which he has answered, as seems to me beyond all doubt, the old vexed question—Whether is the vibration *perpendicular to*, or *in* the plane of polarisation ? To show you this experiment, instead of using unpolarised light for the exciting beam, as in the previous experiment, and holding a small Nicol's prism in my hand and telling you what I saw when I looked through it, I place, as is now done, this great Nicol's prism in the course

of the beam of light before it enters the tube. I
now turn the Nicol's prism into different directions
and turn the apparatus round, so that, sitting in
all parts of the theatre, you may all see the tube
in the proper direction for the successive phenomena of "light," and "no light." You see them
now exactly fulfilling the description which I
gave you in anticipation. If each of you had
a Nicol's prism in your hand, you would learn
that when you see light at all, its plane of polarisation is in the plane through your eye and
the axis of the tube ; and I hope you all now
perfectly understand the proof that the direction
of vibration is perpendicular to this plane.

Now I want to bring before you something
which was taught me a long time ago by Professor Stokes : year after year I have begged him
to publish it, but he has not done so, and so I
have asked him to allow me to speak of it to-night.
It is a dynamical explanation of that wonderful
phenomenon called fluorescence or phosphorescence.
The principle is mechanically represented by this
model (described above with reference to Fig. 34,

p. 157). A simple harmonic motion is, as you now see sustained by my hand in the uppermost bar, in a period of about four seconds. You see that a regular wave-motion travels down the line of molecules represented by these circular disks on the ends of the bars, and the energy continually given to the top bar, by my hand, is continually consumed in heating the basin of treacle and water at the foot. I now remove my hand and leave the whole system to itself. The very considerable sum of kinetic and potential energies of the large masses and spiral springs, attached to the top bar, is gradually spent in sending the diminishing series of waves down the line, and is ultimately converted into heat in the treacle and water. You see that about half of the amplitude of vibration, and therefore three-fourths of the energy, is lost in half a minute.

You will see on quickening the oscillation how very different the result will be. The quick oscillations which I now give to the top bar (the period having been reduced to about one and a half seconds), is incapable of sending waves along the

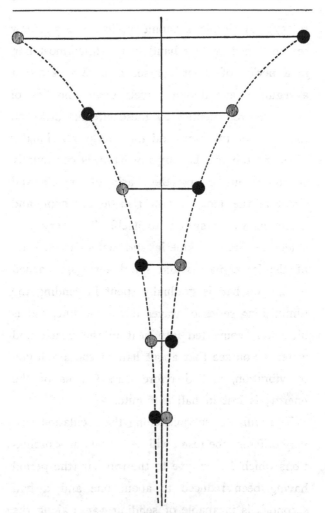

FIG. 42.—Diagram showing the different amplitudes of vibration of a row of particles oscillating in a period less than their least wave-period.

line of molecules ; and it is that rapid oscillation of the particles which, according to Stokes, constitutes latent or stored-up light (see Fig. 42). Remark now that when I remove my hand from the top bar, as no waves travel down the line, no energy is spent in the treacle ; and the vibration goes on for ever (or, to be more exact, say for one minute) as you see, with *no loss* (or, to be quite in accordance with what we see, let me say scarcely any sensible loss). This is a mechanical model, correctly illustrating the dynamical principle of Stokes' explanation of phosphorescence or stored-up light, stored as in the now well-known luminous paint, of which you see the action in this specimen, and in the phosphorescent sulphides of lime in these glass tubes kindly lent by Mr. De La Rue. (Experiment shown.)

Now I will show you Stokes' phenomenon of *fluorescence* in a piece of uranium glass. I hold it in the beam from the electric lamp dispersed by the prism as you see. You see the uranium glass made visible—being illuminated by invisible rays. The rays by which it is illuminated, even before it comes into the visible rays, are manifestly invisible so far as the screen receiving the spectrum is a test of visibility; because the uranium glass, and my hands holding it, throw no shadow on the screen. Also you see the uranium glass which I hold in my hand in the ultra-violet light, while you do not see my hand. I now bring it nearer the place where you see the air (or rather the dust in it) illuminated by the violet light: still no shadow on the screen, but the uranium glass in my hand glowing more brilliantly with its green light of very mixed constitution, consisting of waves of longer periods than that of the ultra-violet, which the incident light, of shorter period than that of violet light, causes the particles of the uranium glass to emit. This light is altogether unpolarised. It was the absolute want of polarisation, and the fact of its periods being all less than those of the exciting light, that led Stokes to distinguish this illumination, which you see in the uranium glass[1] from the mere molecular

[1] The same phenomenon is to be seen splendidly in sulphate of quinine. An interesting experiment may be made by writing on

illumination (always polarised partially if not completely, and always of the same period as that of the exciting light) which we were looking at previously in Dr. Tyndall's experiment.

Stokes gave the name of fluorescence to the glowing with light of larger period than the exciting light, because it is observed in fluor spar, and he wished to avoid all hypothesis in his choice of a name. He pointed out a strong resemblance between it and the old known phenomenon of

a white paper screen, with a finger or a brush dipped in a solution of sulphate of quinine. The marking is quite imperceptible in ordinary light ; but if a prismatic spectrum be thrown on the screen, with the ultra-violet invisible light on the part which had been written on with the sulphate of quinine, the writing is seen glowing brilliantly with a bluish light, and darkness all round. The pheno menon presented by sulphate of quinine and many other vegetable solutions, and some minerals, as, for instance, fluor spar, and various ornamental glasses, as a yellow Bohemian glass, called in commerce "canary glass" (giving a dispersed greenish light), had been discovered by Sir David Brewster (*Transactions*, Royal Society of Edinburgh, 1833, and British Association, Newcastle, 1838), and had been investigated also by Sir John Herschel, and by him called "epipolic dispersion" (*Phil. Trans.* 1845). A complete experimental analysis of the phenomenon, showing precisely what it was that the previous observers had seen, and explaining many singularly mysterious things which they had noticed, was made by Stokes, and described in his paper, "On the Change of Refrangibility of Light" (*Phil. Trans.* May 27, 1852).

P

phosphorescence; but he found some seeming contrasts between the two, which prevented him from concluding fluorescence to be in reality a case of phosphorescence.

In the course of a comparison between the two phenomena (sections 221 to 225 of his 1852 paper), the following statement is given:—"But by far the most striking point of contrast between the two phenomena consists in the apparently instantaneous commencement and cessation of the illumination, in the case of internal dispersion when the active light is admitted and cut off. There is nothing to create the least suspicion of any appreciable duration in the effect. When internal dispersion is exhibited by means of an electric spark, it appears no less momentary than the illumination of the landscape by a flash of lightning. I have not attempted to determine whether any appreciable duration could be made out by means of a revolving mirror." The investigation here suggested has been actually made by Edmund Becquerel, and the question—Is there any appreciable duration in the glow of fluorescence?—has been answered affirmatively by this beautiful and simple little machine before you which he invented for the purpose. The experiment giving the answer is most interesting, and I am sure you will see it with pleasure. The apparatus consists of a flat circular box, with two holes facing one another in the flat sides near the circumference; inside are two disks, carried by a rapidly revolving shaft, by which the holes are alternately shut and opened; one opened when the other is closed, and *vice versa*. A little piece of uranium glass is fixed inside the box between the two holes, and a beam of light from the electric lamp falls upon one of the holes. You look at the other.

Now when I turn the shaft slowly you see nothing. At this instant the light falls on the uranium glass through the open hole far from you, but you see nothing, because the hole next you is shut. Now the hole next you is open, but you see nothing, because the hole next the light is shut, and the uranium glass shows no perceptible afterglow as arising from its previous illumination. This agrees exactly with what you saw when I

P 2

held the large slab of uranium glass in the ultra-violet light of the prismatic spectrum. As long as I held the uranium glass there you saw it glowing; the moment I took it out of the invisible light it ceased to glow. The " moment" of which we were then cognisant may have been the tenth of a second. If the uranium glass had continued to glow sensibly for the twentieth or the fiftieth of a second, it would have seemed to our slow-going sense of vision to cease the moment it was taken out. Now I turn the wheel at such a rate that the hole next you is open about a fiftieth of a second after the uranium glass was bathed in light; still you see nothing. I turn it faster and faster and it now begins to glow, when the hole next you is open about the two-hundredth of a second after the immediately preceding admission of light by the other hole. I turn it faster and faster, and it glows more and more brightly, till now it is glowing like a red coal ; further augmentation of the speed shows, as you see, but little difference in the glow.

Thus it seems that fluorescence is essentially

the same as phosphorescence ; and we may expect
that substances will be found continuously bridging
over the difference of quality between this uranium
glass, which glows only for a few thousandths of a
second, and the luminous sulphides which glow for
hours or days or weeks after the cessation of the
exciting light.

The most decisive and discriminating method of
estimating the size of atoms I have left until my
allotted hour is gone—that founded on the kinetic
theory of gases. Here is a diagram (Fig. 43) of a
crowd of atoms or molecules showing, on a scale
of 1,000,000 to 1, all the molecules of air, of which
the centres may at any instant be in the space of
a square of 1/10,000th of a centimetre side and
1/100,000,000th of a centimetre thick. The side
of the square you see in the diagram is a metre,
and represents 1/10,000th of a centimetre. The
diagram shows just 100 molecules, being 1/10,000th
of the whole number of particles (10^6) in the cube
of 1/10,000th centimetre, or all the molecules in a
slice of 1/10,000th of the thickness of that cube.
Think of a cube filled with particles, like these

glass balls,[1] scattered at random through a space equal to 1,000 times the sum of their volumes.

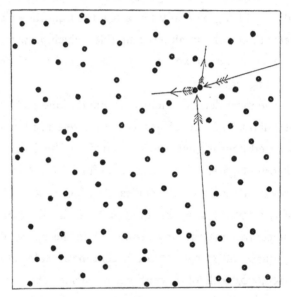

Fig. 43.—Diagram illustrating the number of molecules in a space of 1-10,000th of a centimetre square and 1-100,000,000th of a centimetre thick.

Such a crowd may be condensed (just as air may be condensed) to 1/1000th of its volume, but this

---

[1] The piece of apparatus now exhibited, illustrated the collisions taking place between the molecules of gaseous matter, and the diffusion of one gas into another. It consisted of a board of about

condensation brings the molecules into contact. Something comparable with this may be imagined to be the condition of common air of ordinary density, as in our atmosphere. The diagram, (Fig. 43), with the size of each molecule, such as, if shown in it to scale, would be one millimetre (or too small to be seen by you) to represent an actual diameter of 1/10,000,000th of a centimetre, represents a gas in which a condensation of 1 to 10 linear, or 1 to 1,000 volume would bring the molecules close together.

Now you are to imagine the particles moving in all directions, each in a straight line until it collides with another. The average length of free path is 10 centimetres in our diagram, representing 1/100,000th of a centimetre in

one metre square, perforated with 100 holes in ten rows of ten holes each. From each hole was suspended a cord five metres long. To the lower end of each cord, in five contiguous rows, there was secured a blue coloured glass ball of four centimetres diameter ; and similarly to each cord of the other five rows, a red coloured ball of the same size. A ball from one of the outer rows was pulled aside, and, being set free, it plunged in amongst the others, causing collisions throughout the whole plane in which the suspended balls were situated.

reality. And to suit the case of atmospheric air
of ordinary density and at ordinary pressure you
must suppose the actual velocity of each particle
to be 50,000 centimetres per second, which will
make the average time from collision to collision
1/5,000,000,000th of a second.

The time is so far advanced that I cannot
speak of the details of this exquisite kinetic
theory, but I will just say that three points
investigated by Maxwell and Clausius, viz. the
viscosity or want of perfect fluidity of gases,
the diffusion of gases into one another, and
the diffusion of heat through gases—all these
put together give an estimate for the average
length of the free path of a molecule. Then
a beautiful theory of Clausius enables us,
from the average length of the free path,
to calculate the magnitude of the atom. That
is what Loschmidt has done,[1] and I, un-
consciously following in his wake, have come
to the same conclusion; that is, we have
arrived at the absolute certainty that the

[1] *Sitzungsberichte* of the Vienna Academy, Oct. 12, 1865, p. 395.

dimensions of a molecule of air are something like that which I have stated.

The four lines of argument which I have now indicated lead all to substantially the same estimate of the dimensions of molecular structure. Jointly they establish, with what we cannot but regard as a very high degree of probability, the conclusion that, in any ordinary liquid, transparent solid, or seemingly opaque solid, the mean distance between the centres of contiguous molecules is less than the 1/5,000,000th, and greater than the 1/1,000,000,000th of a centimetre.

To form some conception of the degree of coarse-grainedness indicated by this conclusion, imagine a globe of water or glass, as large as a football,[1] to be magnified up to the size of the earth, each constituent molecule being magnified in the same proportion. The magnified structure would be more coarse-grained than a heap of small shot, but probably less coarse-grained than a heap of footballs.

[1] Or say a globe of 16 centimetres diameter.

STEPS TOWARDS A KINETIC THEORY
OF MATTER.

[*Opening Address to the Mathematical and Physical Section of the British Association, Montreal meeting,* 1884 (*Brit. Assoc. Report,* p. 613).]

THE now well-known kinetic theory of gases is a step so important in the way of explaining seemingly static properties of matter by motion, that it is scarcely possible to help anticipating in idea the arrival at a complete theory of matter, in which all its properties will be seen to be merely attributes of motion. If we are to look for the origin of this idea, we must go back to Democritus, Epicurus and Lucretius. We may then, I believe, without missing a single step, skip 1,800 years. Early last century we find in Malebranche's *Recherche de la Vérité* the statement that *La*

dureté des corps depends on *petits tourbillons.*[1]
These words, embedded in a hopeless mass of
unintelligible statements of the physical, metaphysical, and theological philosophies of the day,
and unsupported by any explanation, elucidation,
or illustration throughout the rest of the three
volumes, and only marred by any other single
sentence or word to be found in the great book,
still do express a distinct conception, which forms
a most remarkable step towards the kinetic theory
of matter. A little later we have Daniel Bernoulli's promulgation of what we now accept as
a surest article of scientific faith—the kinetic
theory of gases. He, so far as I know, thought
only of the Boyle's and Marriot's law of the
"spring of air," as Boyle called it, without reference
to change of temperature or the augmentation of

---

[1] "Preuve de la supposition que j'ay faite : Que la matière subtile
ou éthérée est necessairement composée de PETITS TOURBILLONS ;
et qu'ils sont les causes naturelles de tous changemens qui arrivent
à la matière ; ce que je confirme par l'explication des effets les plus
généraux de la Physique, tels que sont la dureté des corps, leur
fluidité, leur pesanteur, leur légèreté, la lumière réfraction et
réflexion de ses rayons."—MALEBRANCHE, *Recherche de la Vérité,*
1712.

its pressure if not allowed to expand for elevation of temperature, a phenomenon which perhaps he scarcely knew, still less the elevation of temperature produced by compression, and the lowering of temperature by dilatation, and the consequent necessity of waiting for a fraction of a second or a few seconds of time (with apparatus of ordinary experimental magnitude), to see a subsidence from a larger change of pressure, down to the amount of change that verifies Boyle's law. The consideration of these phenomena forty years ago by Joule, in connection with Bernoulli's original conception, formed the foundation of the kinetic theory of gases as now we have it. But what a splendid and useful building has been placed on this foundation by Clausius and Maxwell, and what a beautiful ornament we see on the top of it in the radiometer of Crookes, securely attached to it by the happy discovery of Tait and Dewar,[1] that the length of the free path of the residual molecules of air in a good modern vacuum may amount to several inches! Clausius' and Maxwell's explanations of

[1] *Proc. R. S. E.* March 2, 1874, and July 5, 1875.

the diffusion of gases, and of thermal conduction in gases, their charmingly intelligible conclusion that in gases, the diffusion of heat is just a little more rapid than the diffusion of molecules, because of the interchange of energy in collisions between molecules,[1] while the chief transference of heat is by actual transport of the molecules themselves; and Maxwell's explanation of the viscosity of gases, with the absolute numerical relations which the work of those two great discoverers found

---

[1] On the other hand in liquids, on account of the crowdedness of the molecules, the diffusion of heat must be chiefly by interchange of energies between the molecules, and should be, as experiment proves it is, enormously more rapid than the diffusion of the molecules themselves, and this again ought to be much less rapid than either the material or thermal diffusivities of gases. Thus the diffusivity of common salt through water, was found by Fick to be as small as ·0000116 square centimetres per second : nearly 200 times as great as this is the diffusivity of heat through water, which was found by J. T. Bottomley to be about ·002 square centimetres per second. The material diffusivities of gases, according to Loschmidt's experiments, range from ·098 (the inter-diffusivity of carbonic acid and nitrous oxide) to ·642 (the inter-diffusivity of carbonic oxide and hydrogen) ; while the thermal diffusivities of gases, calculated according to Clausius' and Maxwell's kinetic theory of gases are 089 for carbonic acid, ·16 for common air or other gases of nearly the same density, and 1·12 for hydrogen (all, both material and thermal, being reckoned in square cemtimetres per second).

among the three properties of diffusion, thermal conduction, and viscosity, have annexed to the domain of science a vast and ever-growing province.

Rich as it is in practical results, the kinetic theory of gases, as hitherto developed, stops absolutely short at the atom or molecule, and gives not even a suggestion towards explaining the properties in virtue of which the atoms or molecules mutually influence one another. For some guidance towards a deeper and more comprehensive theory of matter, we may look back with advantage to the end of last century and to the beginning of this century, and find Rumford's conclusion regarding the heat generated in boring a brass gun: "It appears to me to be extremely difficult, if not quite impossible, to form any distinct idea of anything capable of being excited and communicated in the manner the heat was excited and communicated in these experiments, except it be MOTION;"[1] and Davy's still more suggestive statement: "The phenomena of re-

[1] Count Rumford's Works, Vol. I. p. 90: published by the American Academy of Arts and Sciences, Boston, 187 .

pulsion are not dependent on a peculiar elastic fluid for their existence. . . ." " Heat may be defined as a peculiar motion, probably a vibration, of the corpuscles of bodies, tending to separate them. . . ." " To distinguish this motion from others, and to signify the causes of our sensations of heat, &c., the name *repulsive* motion has been adopted."[1] Here we have a most important idea. It would be a somewhat bold figure of speech to say the earth and moon are kept apart by a repulsive motion ; and yet, after all, what is centrifugal force but a repulsive motion, and may it not be that there is no such thing as repulsion, and that it is solely by inertia that what seems to be repulsion is produced ? Two bodies fly together, and, accelerated by mutual attraction, if they do not precisely hit one another, they cannot but separate in virtue of the inertia of their masses. So, after dashing past one another in sharply concave curves round

[1] " Essay on Heat, Light, and the Combinations of Light " ; Collected Works of Sir Humphry Davy, Vol. II. pp. 10, 14, and 20.

their common centre of gravity, they fly asunder again. A careless onlooker might imagine they had repelled one another, and might not notice the difference between what he actually sees and what he would see if the two bodies had been projected with great velocity towards one another, and either colliding and rebounding, or repelling one another into sharply convex continuous curves, fly asunder again.

Joule, Clausius, and Maxwell, and no doubt Daniel Bernoulli himself, and I believe every one who has hitherto written or done anything very explicit in the kinetic theory of gases, has taken the mutual action of molecules in collison as repulsive. May it not after all be attractive? This idea has never left my mind since I first read Davy's *Repulsive Motion*, about thirty-five years ago, but I never made anything of it, at all events have not done so until to-day (June 16, 1884), if this can be said to be making anything of it), when in endeavouring to prepare the present address I notice that Joule's and my own old experiments[1] on the thermal effect of gases expanding from a high pressure vessel through a porous plug, proves the less dense gas to have greater intrinsic *potential* energy than the denser gas, if we assume the ordinary hypothesis regarding the temperature of a gas, according to which two gases are of equal temperatures[2] when the kinetic energies of their constituent molecules are of equal average amounts per molecule.

Think of the thing thus. Imagine a great multitude of particles enclosed by a boundary which may be pushed inwards in any part all round at pleasure. Now station an engineer corps

[1] Republished in Sir W. Thomson's *Mathematical and Physical Papers*, Vol. I. Article XLIX. p. 381 ; also, see Joule's Collected Papers, Vol. II. p. 216.

[2] That this is a mere hypothesis has been scarcely remarked by the founders themselves, nor by almost any writer on the kinetic theory of gases. No one has yet examined the question : what is the condition as regards average distribution of kinetic energy, which is ultimately fulfilled by two portions of gaseous matter, separated by a thin elastic septum which absolutely prevents interdiffusion of matter, while it allows interchange of kinetic energy by collisions against itself? Indeed I do not know but that the present is the very first statement which has ever been published of this condition of the problem of equal temperatures between two gaseous masses.

Q

of Maxwell's army of sorting demons all round the enclosure, with orders to push in the boundary diligently everywhere, when none of the besieged troops are near, and to do nothing when any of them are seen approaching, and until after they have turned again inwards. The result will be that with exactly the same sum of kinetic and potential energies of the same enclosed multitude of particles, the throng has been caused to be denser. Now Joule's and my own old experiments on the efflux of air prove that if the crowd be common air, or oxygen, or nitrogen, or carbonic acid, the temperature is a little higher in the denser than in the rarer condition when the energies are the same. By the hypothesis, equality of temperature between two different gases or two portions of the same gas at different densities means equality of kinetic energies in the same number of molecules of the two. From our observations proving the temperature to be higher, it therefore follows that the potential energy is smaller in the condensed crowd. This —always, however, under protest as to the temperature hypothesis—proves some degree of attraction among the molecules, but it does not prove ultimate attraction between two molecules in collision, or at distances much less than the average mutual distance of nearest neighbours in the multitude. The collisional force might be repulsive, as generally supposed hitherto, and yet attraction might predominate in the whole reckoning of difference between the intrinsic potential energies of the more dense and less dense multitudes. It is, however, remarkable that the explanation of the propagation of sound through gases, and even of the positive fluid pressure of a gas against the sides of the containing vessel, according to the kinetic theory of gases, is quite independent of the question whether the ultimate collisional force is attractive or repulsive. Of course it must be understood that if it is attractive, the particles must be so small that they hardly ever meet—they would have to be infinitely small to *never* meet—that, in fact, they meet so seldom, in comparison with the number of times their courses are turned

through large angles by attraction, that the influence of these purely attractive collisions is
preponderant over that of the comparatively very
rare impacts from actual contact. Thus, after all,
the train of speculation suggested by Davy's
Repulsive Motion does not allow us to escape
from the idea of true repulsion, does not do more
than let us say it is of no consequence, nor even
say this with truth, because, if there are impacts
at all, the nature of the force during the impact,
and the effects of the mutual impacts, however
rare, cannot be evaded in any attempt to realise
a conception of the kinetic theory of gases. And
in fact, unless we are satisfied to imagine the
atoms of a gas as mathematical points endowed
with inertia, and, as according to Boscovich,
endowed with forces of mutual positive and
negative attraction, varying according to some
definite function of the distance, we cannot avoid
the question of impacts, and of vibrations and
rotations of the molecules resulting from impacts,
and we must look distinctly on each molecule
as being either a little elastic solid, or a con-

figuration of motion in a continuous all-pervading liquid. I do not myself see how we can ever permanently rest anywhere short of this last view; but it would be a very pleasant temporary resting-place on the way to it, if we could, as it were, make a mechanical model of a gas out of little pieces of round, perfectly elastic solid matter, flying about through the space occupied by the gas, and colliding with one another and against the sides of the containing vessel. This is, in fact, all we have of kinetic theory of gases up to the present time, and this has done for us, in the hands of Clausius and Maxwell, the great things which constitute our first step towards a molecular theory of matter. Of course from it. we should have to go on to find an explanation of the elasticity and all the other properties of the molecules themselves, a subject vastly more complex and difficult than the gaseous properties for the explanation of which we assume the elastic molecule; but without any explanation of the properties of the molecule itself, with merely the assumption that the mole-

cule has the requisite properties, we might rest
happy for a while in the contemplation of the
kinetic theory of gases, and its explanation of
the gaseous properties, which is not only stupend-
ously important as a step towards a more thorough-
going theory of matter, but is undoubtedly
the expression of a perfectly intelligible and
definite set of facts in nature. But, alas, for our
mechanical model, consisting of the cloud of
little elastic solids flying about amongst one
another. Though each particle have absolutely
perfect elasticity, the end must be pretty much
the same as if it were but imperfectly elastic.
The average effect of repeated and repeated
mutual collisions must be to gradually convert
all the translational energy into energy of shriller
and shriller vibrations of the molecule. It seems
certain that each collision must leave something
more of energy in vibrations of very finely
divided nodal parts than there was of energy
of such vibrations before the impact. The more
minute this nodal subdivision, the less must be
the tendency to give up part of the vibrational

energy into the shape of translational energy in the course of a collision, and I think it rigorously demonstrable that the whole translational energy must ultimately become transformed into vibrational energy of higher and higher nodal subdivisions if each molecule is a continuous elastic solid. Let us, then, leave the kinetic theory of gases for a time with this difficulty unsolved in the hope that we or others after us may return to it, armed with more knowledge of the properties of matter, and with sharper mathematical weapons to cut through the barrier which at present hides from us any view of the molecule itself, and of the effects, other than mere change of translational motion, which it experiences in collision.

To explain the elasticity of a gas was the primary object of the kinetic theory of gases. This object is only attainable by the assumption of an elasticity more complex in character, and more difficult of explanation, than the elasticity of gases—the elasticity of a solid. Thus, even if the fatal fault in the theory, to which I have

alluded, did not exist, and if we could be perfectly satisfied with the kinetic theory of gases
founded on the collisions of elastic solid molecules, there would still be beyond it a grander
theory which need not be considered a chimerical
object of scientific ambition—to explain the
elasticity of solids. But we may be stopped
when we commence to look in the direction of
such a theory with the cynical question: What
do you mean by explaining a property of matter?
As to being stopped by any such question, all
I can say is that if engineering were to be all
and to end all physical science, we should perforce
be content with merely finding properties of
matter by observation, and using them for
practical purposes. But I am sure very few,
if any, engineers are practically satisfied with so
narrow a view of their noble profession. They
must and do patiently observe, and discover by
observation, properties of matter, and results of
material combinations. But deeper questions are
always present, and always fraught with interest
to the true engineer, and he will be the last to

give weight to any other objection to any attempt to see below the surface of things than the practical question : Is it likely to prove wholly futile ? But now instead of imagining the question : What do you mean by explaining a property of matter ? to be put cynically, and letting ourselves be irritated by it, suppose we give to the questioner credit for being sympathetic, and condescend to try and answer his question. We find it not very easy to do so. All the properties of matter are so connected that we can scarcely imagine one *thoroughly explained* without our seeing its relation to all the others, without in fact having the explanation of all, and till we have this we cannot tell what we mean by " explaining a property," or "explaining the properties" of matter. But though this consummation may never be reached by man, the progress of science may be, I believe will be, step by step towards it, on many different roads converging towards it from all sides. The kinetic theory of gases is, as I have said, a true step on one of the roads. On the very distinct road of chemical science, St. Clair Deville arrived at his grand theory of dissociation without the slightest aid from the kinetic theory of gases. The fact that he worked it out solely from chemical observation and experiment, and expounded it to the world without any hypothesis whatever, and seemingly even without consciousness of the beautiful explanation it has in the kinetic theory of gases, secured for it immediately an independent solidity and importance as a chemical theory when he first promulgated it, to which it might even by this time scarcely have attained if it had first been suggested as a probability indicated by the kinetic theory of gases, and been only afterwards confirmed by observation. Now, however, guided by the views which Clausius and Williamson have given us of the continuous interchange of partners between the compound molecules constituting chemical compounds in the gaseous state, we see in Deville's theory of dissociation a point of contact of the most transcendent interest between the chemical and physical lines of scientific progress.

To return to elasticity : if we could make out of matter devoid of elasticity a combined system of relatively moving parts which, in virtue of motion, has the essential characteristics of an elastic body, this would surely be, if not positively a step in the kinetic theory of matter, at least a finger-post pointing a way which we may hope will lead to a kinetic theory of matter. Now this, as I have already shown,[1] we can do in several ways. In the case of the last of the communications referred to, of which only the title has hitherto been published, I showed that, from the mathematical investigation of a gyrostatically dominated combination contained in the passage of Thomson and Tait's *Natural Philosophy* referred to, it follows that any ideal system of material particles, acting on one another mutually through massless connecting springs, may be

[1] Paper on " Vortex Atoms," *Proc. R. S. E.*, Feb. 1867 ; abstract of Lecture before Royal Institution of Great Britain, March 4, 1881, on "Elasticity viewed as possibly a Mode of Motion" (included in the present volume) ; Thomson and Tait's *Natural Philosophy*, second edition, Part I. §§ 345 viii to 345 xxvii ; "On Oscillation and Waves in an Adynamic Gyrostatic System" (title only) *Proc. R. S. E.*, March, 1883.

perfectly imitated in a model consisting of rigid links joined together, and having rapidly rotating fly-wheels pivoted on some or on all of the links. The imitation is not confined to cases of equilibrium. It holds also for vibration produced by disturbing the system infinitesimally from a position of stable equilibrium and leaving it to itself. Thus we may make a gyrostatic system such that it is in equilibrium under the influence of certain positive forces applied to different points of this system; all the forces being precisely the same as, and the points of application similarly situated to, those of the stable system with springs. Then, provided proper masses (that is to say, proper amounts and distributions of inertia) be attributed to the links, we may remove the external forces from each system, and the consequent vibration of the points of application of the forces will be identical. Or we may act upon the systems of material points and springs with any given forces for any given time, and leave it to itself, and do the same thing for the gyrostatic system; the consequent motion will be the same in the two cases. If in the one case the springs are made more and more stiff, and in the other case the angular velocities of the fly-wheels are made greater and greater, the periods of the vibrational constituents of the motion will become shorter and shorter, and the amplitudes smaller and smaller, and the motions will approach more and more nearly those of two perfectly rigid groups of material points moving through space and rotating according to the well-known mode of rotation of a rigid body having unequal moments of inertia about its three principal axes. In one case the ideal nearly rigid connection between the particles is produced by massless exceedingly stiff springs; in the other case it is produced by the exceedingly rapid rotation of the fly-wheels in a system which, when the fly-wheels are deprived of their rotation, is perfectly limp.

The drawings (Figs. 44 and 45) before you llustrate two such material systems.[1] The

In Fig. 44 the two hooked rods seen projecting from the sphere are connected by an elastic coach spring. In Fig. 45 the hooked rods

directions of rotation of the fly-wheels in the gyrostatic system (Fig. 45) are indicated by

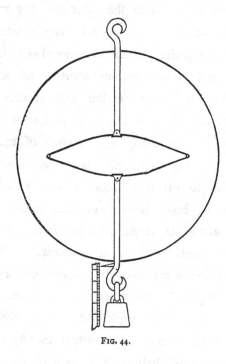

FIG. 44.

directional ellipses, which show in perspective the direction of rotation of the fly-wheel of

are connected one to each of two opposite corners of a four-sided jointed frame, each member of which carries a gyrostat so that the axis of rotation of the fly wheel is in the axis of the member of each gyrostat. The gyrostatic system (Fig. 45)

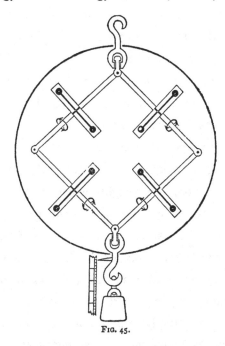

Fig. 45.

might have been constituted of two gyrostatic
members, but four are shown for symmetry.

the frame which bears it. Each of the hooked rods in Fig. 45 is
connected to the framework through a swivel joint, so that the
whole gyrostatic framework may be rotated about the axis of the
hooked rods in order to annul the moment of momentum of the
framework about this axis due to rotation of the fly-wheels in the
gyrostats.

The enclosing circle represents in each case in section an enclosing spherical shell to prevent the interior from being seen. In the inside of one there are fly-wheels, in the inside of the other a massless spring. The projecting hooked rods seem as if they are connected by a spring in each case. If we hang any one of the systems up by the hook on one of its projecting rods, and hang a weight to the hook of the other projecting rod, the weight when first put on will oscillate up and down, and will go on doing so for ever if the system be absolutely unfrictional. If we check the vibration by hand, the weight will hang down at rest, the pin drawn out to a certain degree; and the distance drawn out will be simply proportional to the weight hung on, as in an ordinary spring-balance.

Here, then, out of matter possessing rigidity, but absolutely devoid of elasticity, we have made a perfect model of a spring in the form of a spring-balance. Connect millions of millions of particles by pairs of rods such as these of this spring-balance, and we have a group of particles constituting an elastic solid; exactly fulfilling the mathematical ideal worked out by Navier, Poisson, and Cauchy, and many other mathematicians who, following their example, have endeavoured to found a theory of the elasticity of solids on mutual attraction and repulsion between a group of material particles. All that can possibly be done by this theory, with its assumption of forces acting according to any assumed law of relation to distance, *is* done by the gyrostatic system. But the gyrostatic system does besides, what the system of mutually acting material particles cannot do: it constitutes an elastic solid which can have the Faraday magneto-optic rotation of the plane of polarisation of light; supposing the application of our solid to be a model of the luminiferous ether for illustrating the undulatory theory of light. The gyrostatic model spring-balance is arranged to have zero moment of momentum as a whole, and therefore to contribute nothing to the Faraday rotation; with this arrangement the model illustrates the luminiferous ether in

R

a field unaffected by magnetic force. But now let there be a different rotational velocity imparted to the jointed square round the axis of the two projecting hooked rods, such as to give a resultant moment of momentum round any given line through the centre of inertia of the system, and let pairs of the hooked rods in the model thus altered, which is no longer a model of a mere spring-balance, be applied as connections between millions of pairs of particles as before with the lines of resultant moment of momentum all similarly directed: we now have a model elastic solid which will have the property that the direction of vibration in waves of rectilinear vibrations propagated through it shall turn round the line of propagation of the waves; just as Faraday's observation proves to be done by the line of vibration of light in a dense medium between the poles of a powerful magnet. The case of wave front perpendicular to the lines of resultant moment of momentum (that is to say, the direction of propagation being parallel to these lines) corresponds, in our mechanical model, to the case of light travelling in the direction of the lines of force in a magnetic field.

In these illustrations and models we have different portions of ideal rigid matter acting upon one another, by normal pressure at mathematical points of contact—of course no forces of friction are supposed. It is exceedingly interesting to see how thus, with no other postulates than inertia, rigidity, and mutual impenetrability, we can thoroughly model not only an elastic solid, and any combination of elastic solids, but so complex and recondite a phenomenon as the passage of polarised light through a magnetic field. But now, with the view of ultimately discarding the postulate of rigidity from all our materials, let us suppose some to be absolutely destitute of rigidity, and to possess merely inertia and incompressibility, and mutual impenetrability with reference to the still remaining rigid matter. With these postulates we can produce a perfect model of mutual action at a distance between solid particles, fulfilling the condition, so keenly desired by Newton and Faraday, of being explained by continuous action through an intervening medium. The law of the mutual force in our model, however, is not the simple Newtonian law, but the much more complex law of the mutual action between two electromagnets—with the difference, that in the hydrokinetic model in every case the force is opposite in direction to the corresponding force in the electromagnetic analogue. Imagine a solid bored through with a hole and placed in our ideal perfect liquid. For a moment let the hole be stopped by a diaphragm, and let an impulsive pressure be applied for an instant uniformly over the whole membrane, and then instantly let the membrane be dissolved into liquid. This action originates a motion of the liquid relatively to the solid, of a kind to which I have given the name of "irrotational circulation," which remains absolutely constant however the solid be moved through the liquid. Thus, at any time the actual motion of the liquid at any point in the neighbourhood of the solid will be the resultant of the motion it would have in virtue of the circulation alone, were the solid at rest, and the motion it would have in virtue of the motion of the solid itself, had there been no circulation established through the aperture. It is interesting and important to remark in passing that the whole kinetic energy of the liquid is the sum of the kinetic energies which it would have in the two cases separately. Now, imagine the whole liquid to be enclosed in an infinitely large rigid containing vessel, and in the liquid, at an infinite distance from any part of the containing vessel, let two perforated solids, with irrotational circulation through each, be placed at rest near one another. The resultant fluid motion due to the two circulations will give rise to fluid pressure on the two bodies, which if unbalanced will cause them to move. The force systems—force-and-torques, or pairs of forces—required to prevent them from moving will be mutual and opposite, and will be the same as, but opposite in direction to, the mutual force systems required to hold at rest two electromagnets fulfilling the following specification. The two electromagnets are to be of the same shape and size as the two bodies, and to be placed in the same relative positions, and to consist of infinitely thin layers of electric currents in the surfaces of solids possessing extreme diamagnetic quality—in other words, infinitely small permeability. The distribution of electric current on each body may be any whatever which fulfils the condition that the total current across any closed line drawn on the surface once through the aperture is equal to $1/4\pi$ of the circulation[1] through the aperture in the hydrokinetic analogue.

It might be imagined that the action at a distance thus provided for by fluid motion could serve as a foundation for a theory of the equilibrium, and the vibrations, of elastic solids, and the transmission of waves like those of light

---

[1] The integral of tangential component velocity all round any closed curve, passing once through the aperture, is defined as the "Cyclic constant," or the "circulation" *Vortex Motion*, § 60 *a* (*Trans. R. S. E.*, April 29, 1867). It has the same value for all closed curves passing just once through the aperture, and it remains constant through all time whether the solid body be in motion or at rest.

through an extended quasi-elastic solid medium. But unfortunately for this idea the equilibrium is essentially unstable, both in the case of magnets and, notwithstanding the fact that the forces are oppositely directed, in the hydrokinetic analogue also, when the several movable bodies (two or any greater number) are so placed relatively as to be in equilibrium. If, however, we connect the perforated bodies with circulation through them in the hydrokinetic system, by jointed rigid connecting links, we may arrange for configurations of stable equilibrium. Thus without fly-wheels, but with fluid circulations through apertures, we may make a model spring-balance, or a model luminiferous ether, either without or with the rotational quality corresponding to that of the true luminiferous ether in the magnetic fluid—in short, do all by the perforated solids with circulations through them that we saw we could do by means of linked gyrostats. But something that we cannot do by linked gyrostats we can do by the perforated bodies with fluid circulation. We can make a model gas. The mutual action at a distance, repulsive or attractive according to the mutual aspect of the two bodies when passing within collisional distance [1] of one another, suffices to produce the change of direction of motion in collision, which essentially constitutes the foundation of the kinetic theory of gases; and which, as we have seen before, may as well be due to attraction as to repulsion, so far as we know from any investigation hitherto made in this theory.

There remains, however, as we have seen before, the difficulty of providing for the case of actual impacts between the solids; which must be done by giving them massless spring-

---

[1] According to this view there is no precise distance, or definite condition respecting the distance, between two molecules at which apparently they come to be in collision, or, when receding from one another, they cease to be in collision. It is convenient, however, in the kinetic theory of gases, to adopt arbitrarily a precise definition of collision, according to which two bodies or particles mutually acting at a distance may be said to be in collision when their mutual action exceeds some definite arbitrarily assigned limit, as, for example, when the radius of curvature of the path of either body is less than a stated fraction (1/100, for instance) of the distance between them.

buffers, or, which amounts to the same thing, attributing to them repulsive forces sufficiently powerful at very short distances to absolutely prevent impacts between solid and solid ; unless we adopt the equally repugnant idea of infinitely small perforated solids, with infinitely great fluid circulations through them. Were it not for this fundamental difficulty, the hydrokinetic model gas would be exceedingly interesting ; and, though we could scarcely adopt it as conceivably a true representation of what gases really are, it might still have some importance as a model configuration of solid and liquid matter, by which without elasticity the elasticity of a true gas might be represented.

But lastly, since the hydrokinetic model gas with perforated solids and fluid circulations through them fails because of the impacts between the solids, let us annul the solids and leave the liquid performing irrotational circulation round vacancy,[1] in the place of the solid cores

[1] Investigations respecting coreless vortices will be found in a paper by the author, " Vibrations of a Columnar Vortex," *Proc. R. S. E.*,

which we have hitherto supposed; or let us annul the rigidity of the solid cores of the rings, and give them molecular rotation according to Helmholtz's theory of vortex motion. For stability the molecular rotation must be such as to give the same velocity at the boundary of the rotational fluid core as that of the irrotationally circulating liquid in contact with it, because, as I have proved, frictional slip between two portions of liquid in contact is inconsistent with stability. There is a further condition, upon which I cannot enter in detail just now, but which may be understood in a general way when I say that it is a condition of either uniform or of increasing molecular rotation from the surface inwards, analogous to the condition that the density of a liquid, resting for example under the influence of gravity, must either be uniform or must be greater below than above for stability of equilibrium. All that I have said in favour of

March 1, 1880; and a paper by Hicks, "On the Steady Motion of a Hollow Vortex," read before the Royal Society, June 21, 1883 (see *Trans. R. S.*, for 1884).

the model vortex gas composed of perforated solids with fluid circulations through them holds without modification for the purely hydrokinetic model, composed of either Helmholtz cored vortex rings or of coreless vortices, and we are now troubled with no such difficulty as that of the impacts between solids. Whether, however, when the vortex theory of gases is thoroughly worked out, it will or will not be found to fail in a manner analogous to the failure which I have already pointed out in connection with the kinetic theory of gases composed of little elastic solid molecules, I cannot at present undertake to speak with certainty. It seems to me most probable that the vortex theory cannot fail in any such way, because all I have been able to find out hitherto regarding the vibration of vortices,[1] whether cored or coreless, does not seem to imply the liability of translational or

---

[1] See papers by the author "On Vortex Motion," *Trans. R. S. E.*, April, 1867, and "Vortex Statics," *Proc. R. S. E.*, December, 1875; also a paper by J. J. Thomson, B.A., "On the Vibrations of a Vortex Ring," *Trans. R. S.*, December, 1881, and his valuable book *On Vortex Motion* (being the Adams prize essay for 1882).

impulsive energies of the individual vortices becoming lost in energy of smaller and smaller vibrations.

As a step towards a kinetic theory of matter it is certainly most interesting to remark that in the quasi-elasticity, elasticity looking like that of an india-rubber band, which we see in a vibrating smoke-ring launched from an elliptic aperture, or in two smoke-rings which were circular, but which have become deformed from circularity by mutual collision, we have in reality a virtual elasticity in matter devoid of elasticity and even devoid of rigidity, the virtual elasticity being due to motion, and generated by the generation of motion.

THE SIX GATEWAYS OF KNOWLEDGE.

[*Presidential Address to the Birmingham and Midland Institute, delivered in the Town Hall, Birmingham, on October 3rd, 1883.*]

THE title of the subject upon which I am going to speak this evening, might be—if I were asked to give it a title—"The Six Gateways of Knowledge." I feel that the subject I am about to bring before you, is closely connected with the studies for which the several prizes have been given. The question I will ask you to think of is: What are the means by which the human mind acquires knowledge of external matter?

John Bunyan likens the human soul to a citadel on a hill, self contained, having no means of communication with the outer world, except by five gates—Eye Gate, Ear Gate, Mouth Gate,

Nose Gate, and Feel Gate. Bunyan clearly was in want of a word here. He uses "feel" in the sense of "touch"; a designation which to this day is so commonly used, that I can scarcely accuse it of being incorrect. At the same time, the more correct and distinct designation undoubtedly is, the sense of touch. The late Dr. George Wilson, first Professor of Technology in the University of Edinburgh, published, some time before his death, a beautiful little book under the title of *The Five Gateways of Knowledge;* in which he quotes John Bunyan, in the manner I have indicated to you. But I have said *six* gateways of knowledge, and I must endeavour to justify this saying. I am going to try to prove to you, that we have six senses—that if we are to number the senses at all, we must make them six.

The only census of the senses, so far as I am aware, that ever before made them more than five, was the Irishman's reckoning of seven senses. I presume the Irishman's seventh sense was common sense; and I believe that the possession of that virtue by my countrymen—I speak as an Irishman

—I say the large possession of the seventh sense, which I believe Irishmen have, and the exercise of it, will do more to alleviate the woes of Ireland, than even the removal of the "melancholy ocean" which surrounds its shores. Still, I cannot scientifically see how we can make more than six senses. I shall however, should time permit, return to this question of a seventh sense, and I shall endeavour to throw out suggestions towards answering the question—Is there, or is there not, a magnetic sense? It is possible that there is, but facts and observations so far, give us no evidence that there is a magnetic sense.

The six senses that I intend to explain, so far as I can, this evening, are according to the ordinary enumeration, the sense of sight, the sense of hearing, the sense of smell, the sense of taste, and the sense of touch divided into two departments. A hundred years ago, Dr. Thomas Reid, Professor of Moral Philosophy in the University of Glasgow, pointed out that there was a broad distinction between the sense of roughness or of resistance, which was possessed by the hand, and the sense of heat. Reid's idea has not I think been carried out so much as it deserves. We do not, I believe, find in any of the elementary treatises on Natural Philosophy, or in the physiologists' writings upon the senses, a distinct reckoning of six senses. We have a great deal of explanation about the muscular sense, and the tactile sense ; but we have not a clear and broad distinction of the sense of touch into two departments, which seems to me to follow from Dr. Thomas Reid's way of explaining the sense of touch, although he does not himself distinctly formulate the distinction I am now going to explain.

The sense of touch, of which the organ commonly considered is the hand, but which is possessed by the whole sensitive surface of the body, is very distinctly a double quality. If I touch any object, I perceive a complication of sensations. I perceive a certain sense of roughness, but I also perceive a very distinct sensation, which is not of roughness, or of smoothness. There are two sensations here, let us try to analyse

them. Let me dip my hand into this bowl of
hot water. The moment I touch the water, I
perceive a very distinct sensation, a sensation of
heat. Is that a sensation of roughness, or of
smoothness? No. Again, I dip my hand into
this basin of iced water. I perceive a very distinct
sensation. Is this a sensation of roughness, or of
smoothness? No. Is this comparable with that
former sensation of heat? I say yes. Although
it is opposite, it is comparable with the sensation
of heat. I am not going to say that we have two
sensations in this department; a sensation of heat,
and a sensation of cold. I shall endeavour to
explain that the perceptions of heat and cold
are perceptions of different degrees of one and
the same quality, but that that quality is markedly
different from the sense of roughness. Well now,
what is this sense of roughness? It will take me
some time to explain it fully. I shall therefore
say in advance, that it is a sense of force; and I
shall tell you in advance, before I justify completely what I have to say, that the six senses,
regarding which I wish to give some explanation,

S

are, the sense of sight, the sense of hearing, the sense of taste, the sense of smell, the sense of heat, and the sense of force. The sense of force is the sixth sense ; or the senses of heat and of force are the sense of touch divided into two, to complete the census of six senses that I am endeavouring to demonstrate.

Now I have hinted at a possible seventh sense —a magnetic sense—and though out of the line I propose to follow, and although time is precious, and does not permit much of digression, I wish just to remove the idea that I am in any way suggesting anything towards that wretched super-stition of animal magnetism, and table-turning, and spiritualism, and mesmerism, and clairvoyance, and spirit-rapping, of which we have heard so much. There is no seventh sense of the mystic kind. Clairvoyance, and the like, are the result of bad observation chiefly ; somewhat mixed up, however, with the effects of wilful imposture, acting on an innocent trusting mind. But if there is not a distinct magnetic sense, I say it is a very great wonder that there is not.

We all know a little about the mariner's compass, the needle pointing to the North, and so on
but not many of us have gone far into the subject,
and not many of us understand all the recent
discoveries in electro-magnetism. I could wish,
had I the apparatus here, and if you would allow
me, to show you an experiment in magnetism.
If we had before us a powerful magnet, or say
the machine that is giving us this beautiful electric
light by which the hall is illuminated, it, serving
to excite an electro-magnet, would be one part
of our apparatus ; the other part would be a piece
of copper. Suppose then we had this apparatus,
I would show you a very wonderful discovery
made by Faraday, and worked out admirably by
Foucault, an excellent French experimenter. I
have said that one part of this apparatus would
be a piece of copper, but silver would answer as
well. Probably no other metal than copper or
silver—certainly no other one, of all the metals
that are well known, and obtainable for ordinary
experiments—possesses, and no other metal or
substance whether metallic or not, is known to

possess, in anything like the same degree as copper and silver, the quality I am now going to call attention to.

The quality I refer to is "electric conductivity," and the result of that quality, in the experiment I am now going to describe, is that a piece of copper or a piece of silver, let fall between the poles of a magnet, will fall down slowly as if it were falling through mud. I take this body and let it fall. Many of you here will be able to calculate what fraction of a second it takes to fall one foot. If I took this piece of copper, placed it just above the space between the poles of a powerful electro-magnet and let it go, you would see it fall slowly down before you; it would perhaps take a quarter of a minute, to fall a few inches.

This experiment was carried out in a most powerful manner, by Lord Lindsay (now Lord Crawford), assisted by Mr. Cromwell F. Varley. Both of these eminent men desired to investigate the phenomena of mesmerism, which had been called animal magnetism ; and they very earnestly set to work, to make a real physical experiment.

They asked themselves, Is it conceivable, that if a piece of copper can scarcely move through the air between the poles of an electro-magnet, a human being or other living creature placed there, would experience no effect? Lord Lindsay got an enormous electro-magnet made, so large that the head of any person, wishing to try the experiment, could get well between the poles, in a region of excessively powerful magnetic force. What was the result of the experiment? If I were to say *nothing!* I should do it scant justice. The result was marvellous, and the marvel is that nothing was perceived. Your head, in a space through which a piece of copper falls as if through mud, perceives nothing. I say this is a very great wonder; but I do not admit, I do not feel, that the investigation of the subject is completed. I cannot think that that quality of matter in space—magnetisation— which produces such a prodigious effect upon a piece of metal, can be absolutely without any— it is certainly not without any—effect whatever on the matter of a living body; and that it can be absolutely without any *perceptible* effect whatever on the matter of a living body placed there, seems to me not proved even yet, although nothing has been found. It is so marvellous that there should be no effect at all, that I do believe and feel, that the experiment is worth repeating; and that it is worth examining, whether or not an exceedingly powerful magnetic force has any perceptible effect upon a living vegetable or animal body. I spoke then of a seventh sense. I think it just possible that there may be a magnetic sense. I think it possible, that an exceedingly powerful magnetic effect, may produce a sensation that we cannot compare with heat or force, or any other sensation.

Another question that often occurs is, " Is there an electric sense?" Has any human being a perception of electricity in the air? Well, somewhat similar proposals for experiment might, perhaps, be made with reference to electricity; but there are certain reasons, that would take too long for me to explain, that prevent me from placing the electric force at all in the same category with magnetic force. There would be a surface action that would annul practically the action, due to the electric force, in the interior; and this surface action would be a definite sensation which we could distinctly trace to the sense of touch. Any one putting his hand, or his face, or his hair, in the neighbourhood of an electric machine, perceives a sensation, and on examining it he finds that there is a current of air blowing and that his hair is attracted ; and if he puts his hand too near, he finds that there are sparks passing between his hand or face, and the machine; so that before we come to any subtle question of a possible sense of electric force, we have distinct mechanical agencies, which give rise to senses of temperature and force. But that this mysterious wonderful magnetic force, due, as we now know, to rotations of the molecules, could be absolutely without effect—without perceptible effect—on animal economy, seems a very wonderful result, and at all events it is a subject deserving careful investigation. I hope no one will think that I am favouring the superstition of mesmerism in what I have said.

I intend to explain a little more fully our

perceptions in connection with the double sense of touch—the sense of temperature, and the sense of force—should time permit before I conclude. But I must first say something of the other senses, because if I speak too much about the senses of force and heat, no time will be left for any of the others. Well now, let us think what it is we perceive in the sense of hearing. Acoustics is the science of hearing. And what is hearing? Hearing is perceiving something with the ear. What is it we perceive with the ear? It is something we can also perceive without the ear; something that the greatest master of sound, in the poetic and artistic sense of the word at all events, that ever lived—Beethoven—for a great part of his life could not perceive with his ear at all. He was deaf for a great part of his life, and during that period were composed some of his grandest musical compositions, and that without the possibility of his ever hearing them by ear himself; for his hearing by ear was gone from him for ever. But he used to stand with a stick pressed against the piano and touching his teeth, and thus he could hear the sounds that he called forth from the instrument. Hence, besides the Ear Gate of John Bunyan, there is another gate or access for the sense of hearing.

What is it that you perceive ordinarily by the ear—that a healthy person, without the loss of any of his natural organs of sense, perceives with his ear, but which can otherwise be perceived, although not so satisfactorily or completely? It is distinctly a sense of varying pressure. When the barometer rises, the pressure on the ear increases; when the barometer falls, that is an indication that the pressure on the ear is diminishing. Well, if the pressure of air were suddenly to increase and diminish, say in the course of a quarter of a minute—suppose in a quarter of a minute, the barometer rose one-tenth of an inch, and fell again; would you perceive anything? I doubt it; I do not think you would. If the barometer were to rise two inches, or three inches, or four inches, in the course of half a minute, most people would perceive it. I say this as a result of observation, because people going down in a diving bell have exactly the same sensation as they would experience if from some unknown cause the barometer quickly, in the course of half a minute, were to rise five or six inches—far above the greatest height it ever stands at in the open air. Well now, we have a sense of barometric pressure, but we have not a continued indication that allows us to perceive the difference between the high and low barometer. People living at great altitudes—up several thousand feet above the level of the sea, where the barometer stands several inches lower than at sea level—feel very much as they would do at the surface of the sea, so far as any sensation of pressure is concerned. Keen mountain air feels different from air in lower places partly because it is colder and drier, but also because it is less dense, and you must breathe more of it to get the same quantity of oxygen into your lungs, to perform those functions which the students of the Institute who study animal physiology—and I understand there are a large number—will perfectly understand. The effect of the air in the lungs—the function it performs—depends chiefly on the oxygen taken in. If the air has only three quarters of the density it has in our ordinary atmosphere here, then one and one-third times as much must be inhaled, to produce the same oxidising effect on the blood, and the same general effect in the animal economy; and in that way undoubtedly mountain air has a very different effect on living creatures from the air of the plains. This effect is distinctly perceptible in its relation to health.

But I am wandering from my subject, which is the consideration of the changes of pressure comparable with those that produce sound. A diving bell allows us to perceive a sudden increase of pressure, but not by the ordinary sense of touch. The hand does not perceive the difference between 15 lbs. per square inch pressing it all around, and 17 lbs., or 18 lbs., or 20 lbs., or even 30 lbs. per square inch, as is experienced when you go down in a diving bell. If you go down five and a half fathoms in a diving bell, your hand is pressed all round with a force of 30 lbs. to the square inch; but yet you do not perceive any difference in the sense of force, any perception of pressure. What you do perceive is this : behind the tympanum is a certain cavity filled with air, and a greater pressure on one side of the tympanum than on the other, gives rise to a painful sensation, and sometimes produces rupture of it in a person going down in a diving bell suddenly. The remedy for the painful sensation thus experienced, or rather I should say its prevention, is to keep chewing a piece of hard biscuit, or making believe to do so. If you are chewing a hard biscuit, the operation keeps open a certain passage, by which the air pressure getting access to the inside of the tympanum balances the outside pressure and thus prevents the painful effect. This painful effect on the ear experienced by going down in a diving bell, is simply because a certain piece of tissue is being pressed more on one side than on the other ; and when we get such a tremendous force on a delicate thing like the tympanum, we may experience a great deal of pain, and it may be dangerous ; indeed it is dangerous, and produces rupture or damage to the tympanum unless means be adopted for obviating the difference in the pressures; but the simple means I have indicated are, I believe, with all ordinary healthy persons, perfectly successful.

I am afraid we are no nearer, however, to understanding what it is we perceive when we hear. To be short then it is simply this : it is exceedingly sudden changes of pressure acting on the tympanum of the ear, through such a short time and with such moderate force as not to hurt it ; but to give rise to a very distinct sensation, which is communicated through a train of bones to the auditory nerve. I must merely pass over this ; the details are full of interest, but they would occupy us far more than an hour if I entered upon them at all. As soon as we get to the nerves and the bones, we have gone beyond the subject I proposed to speak upon. My subject belongs to physical science ;—what is called in Scotland, Natural Philosophy. Physical science refers to dead matter, and I have gone beyond its range whenever I speak of a living body ; but we must speak of a living body in dealing with the senses as the means of perceiving—as the means by which, in John Bunyan's language, the "soul in its citadel" acquires a knowledge of external matter. The physicist has to think of the organs of sense, merely as he thinks of the microscope; he has nothing to do with physiology. He has a great deal to do with his own eyes and hands, however, and must think of them, if he would understand what he is doing, and wishes to get a reasonable view of the subject, whatever it may be, which is before him in his own department.

Now, what is the external object of this internal action of hearing and perceiving sound? The external object is a change of pressure of air. Well, but how are we to define a sound simply? It looks a little like a vicious circle, but is not really so, to say it is sound if we call it a sound—if we perceive it *as* sound, it *is* sound. Any change of pressure, which is so sudden as to let us perceive it as sound is a sound. There [giving a sudden clap of the hands]—that is a sound. There is no question about it—nobody will ever ask: Is it a sound or not? It is sound if you hear it. If you do not hear it, it is not to you a sound. That is all I can say to define sound. To explain what it is, I can say, it is change of pressure, and it differs from a gradual change of pressure as seen on the barometer only in being more rapid, so rapid that we perceive it as a sound. If you could perceive by the ear that the barometer has fallen two-tenths of an inch to-day that would be sound. But nobody perceives by his ear that the barometer has fallen, and so he does not hear the fall as a sound. But the same difference of pressure coming on us suddenly—a fall of the barometer, if by any means it could happen, amounting to a tenth of an inch, and taking place in a thousandth of a second,—would affect us quite like sound. A sudden rise of the barometer would produce a sound analogous to what happened when I clapped my hands. What is the difference between a noise and a musical sound? Musical sound is a regular and periodic change of pressure. It is an alternate augmentation and diminution of air pressure, occurring rapidly enough to be perceived as a sound, and taking place with perfect regularity, period after period. Noises and musical sounds merge into one another. Musical sounds have a possibility at least of sometimes ending in noise, or tending too much to a noise, to altogether please a fastidious musical ear. All roughness, irregularity, want of regular smooth periodicity, has the effect of playing out of tune, or of music that is so complicated that it is impossible to say whether it is in tune or not.

But now, with reference to this sense of sound, there is something I should like to say as to the practical lesson to be drawn from the great mathematical treatises which were placed before the British Association, in the addresses of its president, Professor Cayley, and of the president of the mathematical and physical section, Professor Henrici. Both of these professors dwelt on the importance of graphical illustration, and one graphical illustration of Professor Cayley's address may be adduced in respect of this very quality of sound. In the language of mathematics we have just "one independent variable" to deal with in sound, and that is air pressure. We have not a complication of motions in various directions. We have not the complication that we shall have to think of presently, in connection with the sense of force ; complication as to the place of application, and the direction, of the force. We have not the infinite complications we have in some of the other senses, notably smell and taste. We have distinctly only one thing to consider, and that is air pressure, or the variation of air pressure. Now when we have one thing that varies, that, in the language of mathematics, is "one independent variable." Do not imagine that mathematics is harsh and crabbed, and repulsive to common sense. It is merely the etherealisation of common sense. The function of one independent variable that you have here to deal with is the pressure of air on the tympanum. Well now, in a thousand counting-houses and business offices in Birmingham and London, and Glasgow and Manchester, a curve, as Professor Cayley pointed out, is regularly used to show to the eye a function of one independent variable. The function of one independent variable most important in Liverpool perhaps may be the price of cotton. A curve showing the price of cotton, rising when the price of cotton is high, and sinking when the price of cotton is low, shows all the complicated changes of that independent variable to the eye. And so in the Registrar-General's tables of mortality, we have curves showing the number of deaths from day to day—the painful history of an epidemic, shown in a rising branch; and the long gradual talus in a falling branch of the curve when the epidemic is overcome, and the normal state of health is again approached. All that is shown to the eye; and one of the most beautiful results of mathematics is the means of showing to the eye the law of variation, however complicated, of one independent variable. But now for what really to me seems a marvel of marvels: think what a complicated thing is the result of an orchestra playing—a hundred in-struments—and two hundred voices singing in chorus accompanied by the orchestra. Think of

the condition of the air, how it is lacerated sometimes in a complicated effect. Think of the
smooth gradual increase and diminution of pressure—smooth and gradual though taking place
several hundred times in a second—when a piece
of beautiful harmony is heard! Whether however it be the single note of the most delicate
sound of a flute, or the purest piece of harmony of
two voices singing perfectly in tune ; or whether
it be the crash of an orchestra, and the high notes,
sometimes even screechings and tearings of the
air, which you may hear fluttering above the
sound of the chorus—think of all that, and yet
that is not too complicated to be represented
by Professor Cayley, with a piece of chalk in
his hand, drawing on the blackboard a single
line. A single curve, drawn in the manner of
the curve of prices of cotton, describes all that
the ear can possibly hear, as the result of the
most complicated musical performance. How is
one sound more complicated than another ? It
is simply that in the complicated sound the
variations of our one independent variable, pres-

276 POPULAR LECTURES AND ADDRESSES.

sure of air, are more abrupt, more sudden, less
smooth, and less distinctly periodic, than they
are in the softer, and purer, and simpler sound.
But the superposition of the different effects is
really a marvel of marvels ; and to think that
all the different effects of all the different instruments can be so represented ! Think of it in
this way. I suppose everybody present knows
what a musical score is—you know, at all events,
what the notes of a hymn tune look like, and
can understand the like for a chorus of voices,
and accompanying orchestra ;—a "score" of a
whole page with a line for each instrument, and
with perhaps four different lines for four voice
parts. Think of how much you have to put down
on a page of manuscript or print, to show what
the different performers are to do. Think, too,
how much more there is to be done, than anything the composer can put on the page. Think
of the expression which each player is able to
give, and of the difference between a great player
on the violin, and a person who simply grinds
successfully through his part ; think, too, of the

difference in singing, and of all the expression
put into a note or a sequence of notes in singing, that cannot be written down. There is, on
the written or printed page, a little wedge showing
a *diminuendo*, and a wedge turned the other
way showing a *crescendo*, and that is all that
the musician can put on paper to mark the
difference of expression which is to be given.
Well now, all that can be represented by a whole
page or two pages of orchestral score, as the
specification of the sound to be produced in, say,
ten seconds of time, is shown to the eye with
perfect clearness by a single curve on a riband
of paper a hundred inches long. That to my
mind is a wonderful proof of the potency of
mathematics. Do not let any student in this
Institute be deterred for a moment from the pur-
suit of mathematical studies by thinking that the
great mathematicians get into the realm of four
dimensions where you cannot follow them. Take
what Professor Cayley, himself, in his admirable
address which I have already referred to, told
us of the beautiful and splendid power of mathe-

matics for etherealising and illustrating common sense, and you need not be disheartened in your study of mathematics, but may rather be re-invigorated when you think of the power which mathematicians, devoting their whole lives to the study of mathematics, have succeeded in giving to that marvellous science.

I spoke of the sense of sound being caused by rapid variations of pressure. I had better particularise, and say how rapid must be the alternations from greatest pressure to least, and back to greatest, and how frequently must that period occur, to give us the sound of a musical note. If the barometer varies once a minute you would not perceive that as a musical note. But suppose by any mechanical action in the air, you could cause the barometric pressure —the air pressure—to vary much more rapidly. That change of pressure which the barometer is not quick enough to show to the eye, the ear hears as a musical sound if the period recurs twenty times per second. If it recurs twenty, thirty, forty, or fifty times per second, you hear a low note. If the period is gradually accelerated you hear the low note gradually rising, becoming higher and higher, more and more acute, and if it gets up to 256 periods per second, we have a certain note called C in the ordinary musical notation. I believe I describe it correctly as the low note C, of the tenor voice—the gravest C that can be made by a flute. The note of a two-foot organ pipe open at both ends has 256 periods per second. Go on higher and higher to 512 periods per second, and you have the C above that—the chief C of the soprano voice. Go above that to 1,024, you get an octave higher. You get an octave higher always by doubling the number of vibrations per second, and if you go on till you get up to about 5,000 or 6,000 or 10,000 periods per second, the note becomes so shrill that it ceases to excite the human ear, and you do not hear it any longer. The highest note that can be perceived by the human ear seems to be something like 10,000 periods per second. I say "something like," because there is no very definite limit. Some ears cease to hear a note becoming shriller and shriller, before other ears cease to hear it; and, therefore, I can only say in a very general way, that something like 10,000 periods per second is about the shrillest note the human ear is adapted to hear. We may define musical notes therefore as changes of pressure of the air, regularly alternating in periods which lie between twenty and 10,000 per second. Well now, are there vibrations of thirty, or forty, or fifty, or a hundred thousand or a million of periods per second in air, in elastic solids, or in any matter affecting our senses? We have no evidence of the existence in matter of vibrations of very much greater frequency than 10,000, or 20,000, or 30,000 per second, yet we have no reason to deny the possibility of such vibrations existing, and having a large function to perform in nature. But when we get to some degree of frequency that I cannot put figures upon, to something that may be measured in hundred-thousands, if not in millions, of vibrations per second, we have not merely passed the limits of the human ear to hear, but we have passed

the limits of matter, as known to us, to vibrate.
Vibrations transmitted as waves through steel, or
air, or water, cannot be more frequent than a
certain number, which I cannot now put a
figure to, but which, I say, may be reckoned
in hundred-thousands or a few millions per
second.

But now let us think of light. The sense of
sight may be compared to the sense of sound in
this respect—that it also is a matter of vibration.
Light we know to be an influence on the retina of
the eye, and through the retina on the optic
nerve ; an influence dependent on vibrations, whose
frequency is something between 400 million millions
per second and 800 million millions per second.
Now we have a vast gap between 400 per second,
the sound of a rather high tenor voice, and 400
million millions per second, the number of vibrations corresponding to dull red light—the gravest
red light of the prismatic spectrum. Take the
middle of the spectrum—yellow light—the period
of the vibrations there is in round numbers 500
million millions per second. In violet light we

have 800 million millions per second. Beyond
that we have something that the eye scarcely
perceives—does not perceive at all perhaps—but
which I believe it does perceive, though not
vividly ; we have the ultra-violet rays, known
to us chiefly by their photographic effect, but
known also by many other wonderful experiments
which within the last thirty years have enlarged
our knowledge of light to a most marvellous
degree. We have invisible rays of light made
visible by letting them fall on a certain kind of
glass, glass tinged with uranium—that yellowish-
green glass, sometimes called canary glass or
chameleon glass. Uranium glass has a property of
rendering visible to us invisible rays. You may
hold a piece of uranium glass in your hand,
illuminated by this electric light, or by a candle
or by gas light, or hold it in the prismatic
spectrum of white light, and you see it glowing
according to the colour of the light which falls
upon it ; but place it in the spectrum, beyond
the visible violet end, where without it you see
nothing, where a piece of chalk held up seems

quite dark, and the uranium glass glows with a mysterious altered colour of a beautiful tint, revealing the presence of invisible rays, by converting them into rays of lower period, and so rendering them visible to the eye. The discovery of this property of uranium glass was made by Professor Stokes, and the name of fluorescence, from fluor spar, which he found to have the same property, was given to it. It has since been discovered that fluorescence and phosphorescence are continuous, being extremes of the same phenomenon. I suppose most persons here present know the luminous paint made from sulphides of calcium and other materials, which, after being steeped in light for a certain time, keep on for hours giving out light in the darkness. Persistence in emission of light after the removal of the source, which is the characteristic of those phosphorescent objects, is manifested also, as Edmund Becquerel has proved, by the uranium glass, and thus Stokes' discovery of fluorescence comes to be continuous with the old known phenomenon of phosphorescence, to which attention

seems to have been first called scientifically by
Robert Boyle about two hundred years ago.

There are other rays which we do not perceive
in any of these ways, but which we do perceive by
our sense of heat : heat rays as they are commonly
called. But in truth all rays that we call light
have heating effect. Radiant heat and light are one
and indivisible. There are not two things, radiant
heat and light : radiant heat is identical with light.
Take a black hot kettle into a dark room, and
look at it. You do not see it. Hold your face
or your hand near it, and you perceive it by what
Bunyan would have called Feel Gate ; only now
we apply the word feeling to other senses as
well as Touch. You perceive it before you touch
it. You perceive it with the back of your hand,
or the front of your hand ; you perceive it with
your face, yes, and with your eye, but you do
not see it. You perceive it, even by your eye,
and still you do not see it. Well, now, must
I justify the assertion that it is not light ? You
say it is not light, and it is not so to you, if you
do not see it. There has been a good deal of

logic-chopping about the words here; we seem to define in a vicious circle. We may begin by defining light—"It is light if you see it as light; it is not light if you do not see it." To save circumlocution, we shall take things in that way. Radiant heat is light if we see it, it is not light if we do not see it. It is not that there are two things; it is that radiant heat has differences of quality. There are qualities of radiant heat that we can see, and if we see them we call them light; there are qualities of radiant heat we cannot see, and if we cannot see them we do not call them light, but still call them radiant heat: and that on the whole seems to me to be the best logic for this subject.

By the by, I don't see Logic among the studies of the Birmingham and Midland Institute. Logic is to language and grammar what mathematics is to common sense; logic is etherealised grammar. I hope the advanced student in grammar and Latin and Greek, who needs logic perhaps as much as, perhaps more than, most students of science and modern languages, will advance to logic, and consider logic as the science of using words, to lead him to know exactly what he means by them when he uses them. More ships have been wrecked through bad logic than by bad seamanship. When the captain writes down in his log—I don't mean a pun here, log has nothing to do with logic—the ship's place is so-and-so, he means that it is the most probable position—the position which, according to previous observations, he thinks is the most probable. After that, supposing no sights of sun or stars or land to be had, careful observation of speed and direction shows, by a simple reckoning (called technically the *dead-reckoning*), where the ship is next day. But sailors too often forget that what they put down in the log was not the ship's place, but what to their then knowledge was the most probable position of the ship, and they keep running on as if it was the true position. They forget the meaning of the very words in which they have made their entry in the log, and through that bad logic more ships have been run on the rocks than by any other carelessness or bad seamanship. It is bad logic that leads to trusting to the dead-reckoning, in running a course at sea ; and it is that bad logic which is the cause of those terribly frequent wrecks ; of steamers, otherwise well conducted, in cloudy but perfectly fine weather running on rocks at the end of a long voyage. To enable you to understand precisely the meaning of your result when you make a note of anything about your own experience or experiments, and to understand precisely the meaning of what you write down, is the province of logic. To arrange your record in such a manner that if you look at it afterwards it will tell you what it is worth, and neither more nor less is practical logic ; and if you exercise that practical logic, you will find benefits that are too obvious if you only think of any scientific or practical subject with which you are familiar.

There is danger then of a bad use of words, and hence of bad reasoning upon them, in speaking of light and radiant heat ; but if we distinctly define light as that which we consciously perceive

as light—without attempting to define consciousness, because we cannot define consciousness any
more than we can define free will—we shall be
safe. There is no question that you see the
thing: if you see it, it is light. Well now, when
is radiant heat light? Radiant heat is light
when its frequency of vibration is between 400
million millions per second and 800 million
millions per second. When its frequency is less
than 400 million millions per second it is not
light; it is invisible "infra-red" radiant heat. When
its frequency is more than 800 million millions per
second it is not light since we cannot see it;
it is invisible ultra-violet radiation, truly radiant
heat, but it is not so commonly called radiant
heat because its heating effect is known rather
theoretically than by sensory perception, or
thermometric or thermoscopic indications. Observations which have been actually made by
Langley and by Abney on radiant heat take us
down about three octaves below violet, and we
may hope to be brought considerably lower still
by future observation. We know at present in all

about four octaves—that is from one to two, two to four, four to eight, eight to sixteen, hundred million millions per second—of radiant heat. One octave of radiant heat is perceptible to the eye as light, the octave from 400 million millions to 800 million millions. I borrow the word octave from music, not in any mystic sense, nor as indicating any relation between harmony of colours and harmony of sound. No relation exists between harmony of sound and harmony of colours. I merely use the word "octave" as a brief expression for any range of frequencies lying within the ratio of one to two. If you double the frequency of a musical note, you raise it an octave : in that sense I use the word for the moment in respect to light, and in no other sense. Well now, think what a tremendous chasm there is between the 100 million millions per second, which is about the gravest note, hitherto discovered, of invisible radiant heat, and the 10,000 per second, the greatest number of vibrations perceptible as sound. This is an unknown province of science :—the investigation of vibrations between those two

U

limits is, perhaps, one of the most promising provinces of science for the future investigator.

In conclusion, I wish to bring before you the idea that all the senses are related to force. The sense of sound we have seen is merely a sense of very rapid changes of air-pressure (which is force) on the drum of the ear. I have passed merely by name over the senses of taste and smell. I may say they are chemical senses. Taste common salt and taste sugar—you tell in a moment the difference, and the perception of that difference is a perception of chemical quality. There is in this perception a subtle molecular influence, due to the touch of the object on the tongue or the palate, and producing a sensation very different from the ordinarily reckoned sense of touch, which, as we have just seen, tells us only of rough-ness, and of temperature. The most subtle of our senses perhaps is sight; next come smell and taste. Professor Stokes recently told me that he would rather look upon taste and smell and sight as being continuous because they are all molecular—they all deal with properties of matter,

not in the gross, but in their molecular actions—
he would rather group those three together, than
he would couple any one of them with any of the
other senses. It is not necessary, however, for us
to reduce all the six senses to one, but I would
just point out that they are all related to force.
Chemical action is a force, tearing molecules apart,
throwing or pushing them together: and our
chemical sense or senses may, therefore, so far
at least, be regarded as concerned with force.
That the senses of smell and taste are related to
one another, seems obvious; and if physiologists
would pardon me, I would suggest that they
might, without impropriety, be regarded as extremes of one sense. This at all events can be
said of them, they can be compared—which cannot be said of any other two senses. You cannot
say that the shape of a cube, or the roughness
of a piece of loaf sugar or sandstone, is comparable
with the temperature of hot water, or is like
the sound of a trumpet; or that the sound of a
trumpet is like scarlet, or like a rocket, or like a
blue-light signal. There is no comparability

between any of these perceptions. But if any one says, "That piece of cinnamon tastes like its smell," I think he will express something of general experience. The smell and the taste of pepper, nutmeg, cloves, cinnamon, vanilla, apples, strawberries, and other articles of food, particularly spices and fruits, have very marked qualities, in which the taste and the smell seem essentially comparable. It does seem to me, although anatomists distinguish between them because the sensory organs concerned are different, and because they have not discovered a continuity between these organs, that we should not be philosophically wrong in saying that smell and taste are extremes of one sense—one kind of perceptivity—a sense of chemical quality materially presented to us.

Now sense of light, and sense of heat, are very different though we cannot define the difference. You perceive the heat of a hot kettle—how? By its radiant heat against the face—that is one way. But there is another way, not by radiant heat, of which I shall speak later. You perceive by vision, but still in virtue of radiant-heat, a hot body ; if illuminated by light, or if hot enough to be self-luminous, red-hot or white-hot, you see it : you can both see a hot body, and perceive it by its heat, otherwise than by seeing it. Take a piece of red-hot cinder with the tongs, or a red-hot poker, and study it ; carry it into a dark room, and look at it. You see it for a certain time ; after a certain time you cease to see it, but you still perceive radiant heat from it. Well now, there is radiant heat perceived by the eye and the face and the hands all the time ; but it is perceived only by the sense of temperature, when the hot body ceases to be red-hot. There is then, to our senses, an absolute distinction in modes of perception between that which is continuous in the external nature of the thing, namely, radiant heat in its visible and invisible varieties. It operates upon our senses in a way that I cannot ask anatomists to admit to be one and the same in both cases. They cannot now at all events, say that there is an absolute continuity between the retina of the eye in its perception of radiant

heat as light, and the skin of the hand in its
perception of radiant heat as heat. We may come
to know more ; it may yet appear that there is
a continuity. Some of Darwin's sublime specula-
tions, may become realities to us; and we may
come to recognise a cultivable retina all over the
body. We have not done that yet, but Darwin's
grand idea occurs as suggesting that there may
be an absolute continuity, between the perception
of radiant heat by the retina of the eye and its
perception by the tissues and nerves concerned in
the mere sense of heat. We must be content in
the meantime, however, to make a distinction
between the senses of light and heat. And
indeed it must be remarked that our sense of heat
is not excited by radiant heat only, while it is
only and essentially radiant heat that gives to the
retina the sense of light. Hold your hand under
a red-hot poker in a dark room : you perceive
it to be hot solely by its radiant heat, and you
see it also by its radiant heat. Now place the
hand over it : you feel more of heat. Now, in
fact, you perceive its heat in three ways—by

contact with the heated air which has ascended from the poker, and by radiant heat felt by your sense of heat, and by radiant heat seen as light (the iron being still red-hot). But the sense of heat is the same throughout, and is a certain effect experienced by the tissue, whether it be caused by radiant heat, or by contact with heated particles of the air.

Lastly, there remains—and I am afraid I have already taxed your patience too long—the sense of force. I have been vehemently attacked for asserting this sixth sense. I need not go into the controversy, nor try to explain to you the ground on which I have been attacked; I could not in fact, because in reading the attack I have not been able to understand it myself. The only tangible ground of objection, perhaps, was that a writer in New York published this theory in 1880. I had quoted Dr. Thomas Reid, without giving a date; his date chances to be 1780 or thereabouts!! But physiologists have very strenuously resisted admitting that the sense of roughness is the same as that muscular sense, which the metaphysicians who followed Dr. Thomas Reid in the University of Glasgow, taught. It was in the University of Glasgow that I learned about the muscular sense, and I have not seen it very distinctly stated elsewhere. What is this "muscular sense"? I press upon the desk before me with my right hand, or I walk forward holding out my hand in the dark and using this means to feel my way, as a blind man does constantly who finds where he is, and guides himself, by the sense of touch. I walk on until I perceive an obstruction by a sense of force in the palm of the hand. How and where do I perceive this sensation? Anatomists will tell you it is felt in the muscles of the arm. Here, then, is a force which I perceive in the muscles of the arm, and the corresponding perceptivity is properly enough called a muscular sense. But now take the tip of your finger and rub a piece of sandstone, or a piece of loaf sugar, or a smooth table. Take a piece of loaf sugar between your finger and thumb, and take a piece of smooth glass between your finger and thumb. You perceive a difference. What is the difference? It is the sense of roughness as distinguished from smoothness. Physiologists and anatomists have used the word "tactile" sense, to designate it. I confess that this does not convey much to my mind. "Tactile" is merely "of or belonging to touch," and in saying we perceive roughness and smoothness by a tactile sense, we are where we were. We are not enlightened by being told that there is a tactile sense as a department of our sense of touch. But I say the thing thought of is a sense of force. We cannot away with it; it is a sense of force, of directions of forces, and of places of application of forces. If the places of application of the forces are the palms of the two hands, we perceive accordingly, and know that we perceive, in the muscles of the arms, effects of large pressures on the palms of the hands. But if the places of application are a hundred little areas on one finger, we still perceive the effect as force. We distinguish between a uniformly distributed force like the force of a piece of smooth glass, and forces distributed over ten or a hundred little areas. And this is the sense of

smoothness and roughness. The sense of roughness is therefore a sense of forces, and of places
of application of forces, just as the sense of
forces in your two hands stretched out, is the
sense of forces in places at a distance of six
feet apart. Whether the places be at a distance
of six feet or at a distance of one-hundredth
of an inch, it is the sense of force, and of
places of application of forces, and of directions
of forces, that we deal with in the sense of
touch as differing from the sense of heat. Now
anatomists and physiologists have a good right
to distinguish between the kind of excitement
of tissue in the finger and in the minute nerves
of the skin and sub-skin of the finger, by which
you perceive roughness and smoothness, in the
one case ; and of the muscles by which you perceive places of application very distant, in the
other. But whether the forces be so near that
anatomists cannot distinguish muscles—cannot
point out muscles resisting forces and balancing
them—because, remember, when you take a piece
of glass in your fingers every bit of pressure at
every ten-thousandth of an inch pressed by the

glass against the finger is a balanced force—or whether they be far asunder and obviously balanced by the muscles of the two arms, the thing perceived is the same in kind. Anatomists do not show us muscles balancing the individual forces experienced by the small areas of the finger itself when we touch a piece of smooth glass, or the individual forces in the scores or hundreds of little areas, experienced when we touch a piece of rough sugar or rough sandstone ; and perhaps it is not by muscles smaller than the muscles of the finger as a whole that the multitudinousness is dealt with; or perhaps, on the other hand, these nerves and tissues are continuous in their qualities with muscles. I go beyond the range of my subject whenever I speak of muscles and nerves ; but externally the sense of touch other than heat is the same in all cases— it is a sense of forces and of places of application of forces and of directions of forces. I hope now I have justified the sixth sense ; and that I have not taxed your patience unduly in not having done it in fewer words.

THE WAVE THEORY OF LIGHT.

[*A Lecture delivered at the Academy of Music, Philadelphia, under the auspices of the Franklin Institute, September 29th,* 1884.]

THE subject upon which I am to speak to you this evening is happily for me not new in Philadelphia. The beautiful lectures on light which were given several years ago by President Morton, of the Stevens' Institute, and the succession of lectures on the same subject so admirably illustrated by Professor Tyndall, which many now present have heard, have fully prepared you for anything I can tell you this evening in respect to the wave theory of light.

It is indeed my humble part to bring before you only some mathematical and dynamical details of this great theory. I cannot have the

pleasure of illustrating them to you by anything
comparable with the splendid and instructive
experiments which many of you have already
seen. It is satisfactory to me to know that so
many of you, now present, are so thoroughly
prepared to understand anything I can say, that
those who have seen the experiments will not
feel their absence at this time. At the same time
I wish to make them intelligible to those who have
not had the advantages to be gained by a
systematic course of lectures. I must say, in the
first place, without further preface, as time is
short and the subject is long, simply that sound
and light are both due to vibrations propagated
in the manner of waves; and I shall endeavour
in the first place to define the manner of propagation and the mode of motion that constitute
those two subjects of our senses, the sense of sound
and the sense of light.

Each is due to vibrations, but the vibrations
of light differ widely from the vibrations of sound.
Something that I can tell you more easily than
anything in the way of dynamics or mathematics

respecting the two classes of vibrations is, that there is a great difference in the frequency of the vibrations of light when compared with the frequency of the vibrations of sound. The term "frequency" applied to vibrations is a convenient term, applied by Lord Rayleigh in his book on sound to a definite number of full vibrations of a vibrating body per unit of time. Consider, then, in respect to sound, the frequency of the vibrations of notes, which you all know in music represented by letters, and by the syllables for singing, the do, re, mi, &c. The notes of the modern scale correspond to different frequencies of vibrations. A certain note and the octave above it, correspond to a certain number of vibrations per second, and double that number.

I may conveniently explain in the first place the note called 'C'; I mean the middle 'C'; I believe it is the C of the tenor voice, that most nearly approaches the tones used in speaking. That note corresponds to two hundred and fifty-six full vibrations per second—two hundred and fifty-six times to and fro per second of time.

Think of one vibration per second of time. The seconds pendulum of the clock performs one vibration in two seconds, or a half vibration in one direction per second. Take a ten-inch pendulum of a drawing-room clock, which vibrates twice as fast as the pendulum of an ordinary eight-day clock, and it gives a vibration of one per second, a full period of one per second to and fro. Now think of three vibrations per second. I can move my hand three times per second easily, and by a violent effort I can move it to and fro five times per second. With four times as great force, if I could apply it, I could move it twice five times per second.

Let us think, then, of an exceedingly muscular arm that would cause it to vibrate ten times per second, that is, ten times to the left and ten times to the right. Think of twice ten times, that is, twenty times per second, which would require four times as much force; three times ten, or thirty times a second, would require nine times as much force. If a person were nine times as strong as the most muscular arm can be, he could vibrate his hand to and fro thirty times per second, and without any other musical instrument could make a musical note by the movement of his hand which would correspond to one of the pedal notes of an organ.

If you want to know the length of a pedal pipe, you can calculate it in this way. There are some numbers you must remember, and one of them is this. You, in this country, are subjected to the British insularity in weights and measures; you use the foot and inch and yard. I am obliged to use that system, but I apologise to you for doing so, because it is so inconvenient, and I hope all Americans will do everything in their power to introduce the French metrical system. I hope the evil action performed by an English minister whose name I need not mention, because I do not wish to throw obloquy on any one, may be remedied. He abrogated a useful rule, which for a short time was followed, and which I hope will soon be again enjoined, that the French metrical system be taught in all our national schools. I do not know how it is in America. The school system seems to be very admirable, and I hope the teaching of the metrical system will not be let slip in the American schools any more than the use of the globes. I say this seriously: I do not think any one knows how seriously I speak of it. I look upon our English system as a wickedly brain-destroying piece of bondage under which we suffer. The reason why we continue to use it is the imaginary difficulty of making a change, and nothing else; but I do not think in America that any such difficulty should stand in the way of adopting so splendidly useful a reform.

I know the velocity of sound in feet per second. If I remember rightly, it is 1089 feet per second in dry air at the freezing temperature, and 1115 feet per second in air of what we would call moderate temperature, 59 or 60 degrees—(I do not know whether that temperature is ever attained in Philadelphia or not; I have had no experience of it, but people tell me it is sometimes 59 or 60 degrees in Philadelphia, and I believe them)—in round numbers let us call the

x

speed 1000 feet per second. Sometimes we call it a thousand musical feet per second, it saves trouble in calculating the length of organ pipes ; the time of vibration in an organ pipe is the time it takes a vibration to run from one end to the other and back. In an organ pipe 500 feet long the period would be one per second ; in an organ pipe ten feet long the period would be 50 per second ; in an organ pipe twenty feet long the period would be 25 per second at the same rate. Thus 25 per second, and 50 per second of frequencies correspond to the periods of organ pipes of 20 feet and 10 feet.

The period of vibration of an organ pipe, open at both ends, is approximately the time it takes sound to travel from one end to the other and back. You remember that the velocity in dry air in a pipe 10 feet long is a little more than 50 periods per second ; going up to 256 periods per second, the vibrations correspond to those of a pipe two feet long. Let us take 512 periods per second ; that corresponds to a pipe about a foot long. In a flute, open at both ends, the holes are

so arranged that the length of the sound-wave
is about one foot, for one of the chief "open
notes." Higher musical notes correspond to greater
and greater frequency of vibration, viz., 1,000,
2,000, 4,000 vibrations per second; 4,000 vibrations per second correspond to a piccolo flute of
exceedingly small length; it would be but one and
a half inches long. Think of a note from a little
dog-call, or other whistle, one and a half inches
long, open at both ends, or from a little key
having a tube three quarters of an inch long, closed
at one end; you will then have 4,000 vibrations
per second.

A wave length of sound is the distance traversed
in the period of vibration. I will illustrate what
the vibrations of sound are by this condensation
travelling along our picture on the screen. Alternate condensations and rarefactions of the air are
made continuously by a sounding body. When
I pass my hand vigorously in one direction, the
air before it becomes dense, and the air on the
other side becomes rarefied. When I move it in the
other direction these things become reversed; there

X 2

is a spreading out of condensation from the place where my hand moves in one direction and then in the reverse. Each condensation is succeeded by a rarefaction. Rarefaction succeeds condensation at an interval of one-half what we call "wavelengths." Condensation succeeds condensation at the full interval of a wave-length.

We have here these luminous particles on this scale,[1] representing portions of air close together, more dense ; a little higher up, portions of air less dense. I now slowly turn the handle of the apparatus in the lantern, and you see the luminous sectors showing condensation travelling slowly upwards on the screen ; now you have another condensation making one wave-length.

This picture or chart represents a wave-length of four feet. It represents a wave of sound four feet long. The fourth part of a thousand is 250. What we see now of the scale represents the lower note C of the tenor voice. The air from the mouth of a singer is alternately condensed and rarefied

[1] Alluding to a moving diagram of wave motion of sound produced by a working slide for lantern projection.

just as you see here. But that process shoots
forward at the rate of about one thousand feet
per second; the exact period of the motion
being 256 vibrations per second for the actual
case before you.

Follow one particle of the air forming part of
a sound wave, as represented by these moving
spots of light on the screen; now it goes down,
then another portion goes down rapidly; now
it stops going down; now it begins to go up;
now it goes down and up again. As the
maximum of condensation is approached it is
going up with diminishing maximum velocity
The maximum of rarefaction has now reached
it, and the particle stops going up and begins
to move down. When it is of mean density
the particles are moving with maximum velocity,
one way or the other. You can easily follow these
motions, and you will see that each particle moves
to and fro and the thing that we call *condensation*
travels along.

I shall show the distinction between these
vibrations and the vibrations of light. Here is the

fixed appearance of the particles when displaced but not in motion. You can imagine particles of something, the thing whose motion constitutes light. This thing we call the luminiferous ether. That is the only substance we are confident of in dynamics. One thing we are sure of, and that is the reality and substantiality of the luminiferous ether. This instrument is merely a method of giving motion to a diagram designed for the purpose of illustrating wave motion of light. I will show you the same thing in a fixed diagram, but this arrangement shows the mode of motion.

Now follow the motion of each particle. This represents a particle of the luminiferous ether, moving at the greatest speed when it is at the middle position.

You see the two modes of vibration,[1] sound and light now moving together ; the travelling of the wave of condensation and rarefaction, and the travelling of the wave of transverse displacement. Note the direction of propagation. Here it is from

[1] Showing two moving diagrams, simultaneously, on the screen, one depicting a wave motion of light, the other a sound vibration.

your left to your right, as you look at it. Look at the motion when made faster. We have now the direction reversed. The propagation of the wave is from right to left, again the propagation of the wave is from left to right ; each particle moves perpendicularly to the line of propagation.

I have given you an illustration of the vibration of sound waves, but I must tell you that the movement illustrating the condensation and rarefaction represented in that moving diagram are necessarily very much exaggerated, to let the motion be perceptible, whereas the greatest condensation in actual sound motion is not more than one or two per cent. or a small fraction of a per cent. Except that the amount of condensation was exaggerated in the diagram for sound, you have in the chart a correct representation of what actually takes place in sounding the low note C.

On the other hand, in the moving diagram representing light waves what had we? We had a great exaggeration of the inclination of the line of particles. You must first imagine a line of particles in a straight line, and then you must imagine them disturbed into a wave-curve, the shape of the curve corresponding to the disturbance. Having seen what the propagation of the wave is, look at this diagram and then look at that one. This, in light, corresponds to the different sounds I spoke of at first. The wave-length of light is the distance from crest to crest of the

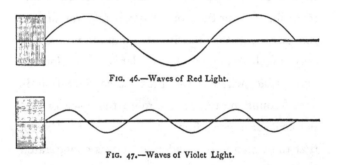

FIG. 46.—Waves of Red Light.

FIG. 47.—Waves of Violet Light.

wave, or from hollow to hollow. I speak of crests and hollows, because we have a diagram of ups and downs as the diagram is placed.

Here, then, you have a wave-length.[1] In this lower diagram (Fig. 47) you have a wave-length of

[1] Exhibiting a large drawing, or chart, representing a red and a violet wave of light (reproduced in Figs. 46 and 47).

violet light. It is but one-half the length of the upper wave of red light ; the period of vibration is but half as long. Now there, on an enormous scale, exaggerated not only as to slope, but immensely magnified as to wave-length, we have an illustration of the waves of violet light. The drawing marked "red" (Fig. 46) corresponds to red light, and this lower diagram corresponds to violet light. The upper curve really corresponds to something a little below the red ray of light in the spectrum, and the lower curve to something beyond the violet light. The variation in wave-length between the most extreme rays is in the proportion of four and a half of red to eight of the violet, instead of four and eight ; the red waves are nearly as one to two of the violet.

To make a comparison between the number of vibrations for each wave of sound and the number of vibrations constituting light waves, I may say that 30 vibrations per second is about the smallest number which will produce a musical sound ; 50 per second gives one of the grave pedal notes of an organ, 100 or 200 per second give the low notes of the bass voice, higher notes with 250 per second, 300 per second, 1,000, 4,000 up to 8,000 per second give about the shrillest notes audible to the human ear.

Instead of the numbers, which we have, say in the most commonly used part of the musical scale, *i.e.*, from 200 or 300 to 600 or 700 per second, we have millions of millions of vibrations per second in light waves: that is to say, 400 per second, instead of 400 million million per second, which is the number of vibrations performed when we have red light produced.

An exhibition of red light travelling through space from the remotest star is due to propagation by waves or vibrations, in which each individual particle of the transmitting medium vibrates to and fro 400 million million times in a second.

Some people say they cannot understand a million million. Those people cannot understand that twice two makes four. That is the way I put it to people who talk to me about the incomprehensibility of such large numbers. I say *finitude* is incomprehensible, the infinite in the universe *is* comprehensible.   Now apply a little logic to this.   Is the negation of infinitude incomprehensible?   What would you think of a universe in which you could travel one, ten, or a thousand miles, or even to California, and then find it come to an end?   Can you suppose an end of matter or an end of space?   The idea is incomprehensible.   Even if you were to go millions and millions of miles the idea of coming to an end is incomprehensible.   You can understand one thousand per second as easily as you can understand one per second.   You can go from one to ten, and ten times ten and then to a thousand without taxing your understanding, and then you can go on to a thousand million and a million million.   You can all understand it.

Now 400 million million vibrations per second is the kind of thing that exists as a factor in the illumination by red light.   Violet light, after what we have seen and have had illustrated by that curve (Fig. 47), I need not tell you corresponds to vibrations of about 800 million million per second. There are recognisable qualities of light caused by vibrations of much greater frequency and much less frequency than this. You may imagine vibrations having about twice the frequency of violet light, and others having about one-fifteenth the frequency of red light and still you do not pass the limit of the range of continuous phenomena only a part of which constitutes *visible* light.

When we go below visible red light what have we? We have something we do not see with the eye, something that the ordinary photographer does not bring out on his photographically sensitive plates. It is light, but we do not see it. It is something so closely continuous with *visible* light, that we may define it by the name of *invisible* light. It is commonly called radiant heat; invisible radiant heat. Perhaps, in this thorny path of logic, with hard words flying in our faces, the least troublesome way of speaking of it is to call it radiant heat. The heat effect you experience when you go near a bright hot coal fire, or a hot steam boiler; or when you go

near, but not over, a set of hot water pipes used
for heating a house; the thing we perceive in
our faces and hands when we go near a boiling
pot and hold the hand on a level with it, is
radiant heat; the heat of the hands and face caused
by a hot fire, or by a hot kettle when held *under*
the kettle, is also radiant heat.

You might readily make the experiment with
an earthen teapot; it radiates heat better than
polished silver. Hold your hands below the teapot
and you perceive a sense of heat; above it you
get more heat; either way you perceive heat. If
held over the teapot you readily understand that
there is a little current of hot air rising; if you
put your hand under the teapot you find cold air
rising, and the upper side of your hand is heated
by radiation while the lower side is fanned and
is actually cooled by virtue of the heated kettle
above it.

That perception by the sense of heat, is the
perception of something actually continuous with
light. We have knowledge of rays of radiant heat
perceptible down to (in round numbers) about

four times the wave-length, or one-fourth the period, of visible or red light. Let us take red light at 400 million million vibrations per second, then the lowest radiant heat, as yet investigated, is about 100 million million per second of frequency of vibration.

I had hoped to be able to give you a lower figure. Professor Langley has made splendid experiments on the top of Mount Whitney, at the height of 15,000 feet above the sea-level, with his "Bolometer," and has made actual measurements of the wave length of radiant heat down to exceedingly low figures. I will read you one of the figures; I have not got it by heart yet, because I am expecting more from him.[1] I learned a year and a half ago that the lowest radiant heat observed by the diffraction method of Professor Langley

[1] Since my lecture I have heard from Professor Langley that he has measured the refrangibility by a rock salt prism, and inferred the wave-length of heat rays from a "Leslie cube" (a metal vessel filled with hot water and radiating heat from a blackened side). The greatest wave-length he has thus found is one-thousandth of a centimetre, which is seventeen times that of sodium light—the corresponding period being about thirty million million per second.

November, 1884.—W. T.

corresponds to 28 one hundred thousandths of a centimentre for wave-length, 28 as compared with red light, which is 7·3 ; or nearly four-fold. Thus wave-lengths of four times the amplitude, or one-fourth the frequency per second of red light have been experimented on by Professor Langley and recognised as radiant heat.

Everybody knows the "photographer's light," and has heard of *invisible* light producing visible effects upon the chemically prepared plate in the camera. Speaking in round numbers, I may say that, in going up to about twice the frequency I have mentioned for violet light you have gone to the extreme end of the range of known light of the highest rates of vibration ; I mean to say that you have reached the greatest frequency that has yet been observed. Photographic, or actinic light, as far as our knowledge extends at present, takes us to a little less than one-half the wave length of violet light.

You will thus see that while our acquaintance with wave motion below the red extends down to one quarter of the slowest rate which affects the eye, our knowledge of vibrations at the other end of the scale only comprehends those having twice the frequency of violet light. In round numbers we have 4 octaves of light, corresponding to 4 octaves of sound in music. In music the octave has a range to a note of double frequency. In light we have one octave of visible light, one octave above the visible range and two octaves below the visible range. We have 100 per second, 200 per second, 400 per second (million million understood) for invisible radiant heat; 800 per second for visible light, and 1,600 per second for invisible or actinic light.

One thing common to the whole is the heat effect. It is extremely small in moonlight, so small that until recently nobody knew there was any heat in the moon's rays. Herschel thought it was perceptible in our atmosphere by noticing that it dissolved away very light clouds, an effect which seemed to show in full moonlight more than when we have less than full moon. Herschel, however, pointed this out as doubtful; but now, instead of its being a doubtful question, we have

Professor Langley giving as a fact that the light from the moon drives the indicator of his sensitive instrument clear across the scale, showing a comparatively prodigious heating effect!

I must tell you that if any of you want to experiment with the heat of the moonlight, you must measure the heat by means of apparatus which comes within the influence of the moon's rays only. This is a very necessary precaution; if, for instance, you should take your Bolometer or other heat detector from a comparatively warm room into the night air, you would obtain an indication of a fall in temperature owing to this change. You must be sure that your apparatus is in thermal equilibrium with the surrounding air, then take your burning-glass, and first point it to the moon and then to space in the sky beside the moon; you thus get a differential measurement in which you compare the radiation of the moon with the radiation of the sky. You will then see that the moon has a distinctly heating effect.

To continue our study of visible light, that is

Y

undulations extending from red to violet in the spectrum (which I am going to show you presently), I would first point out on this chart (Fig. 48) that in the section from letter *A* to letter *D* we have visual effect and heating effect only; but no ordinary chemical or photographic effect. Photographers can leave their usual sensitive chemically prepared plates exposed to yellow light and red light without experiencing

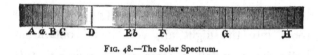

FIG. 48.—The Solar Spectrum.

any sensible effect; but when you get toward the blue end of the spectrum the photographic effect begins to tell, and more and more strongly as you get towards the violet end. When you get beyond the violet there is the invisible light known chiefly by its chemical action. From yellow to violet we have visual effect, heating effect, and chemical effect, all three; above the violet only chemical and heating effects, and so little of the heating effect that it is scarcely perceptible.

The prismatic spectrum is Newton's discovery of the composition of white light. White light consists of every variety of colour from red to violet. Here, now, we have Newton's prismatic spectrum, produced by a prism. I will illustrate a little in regard to the nature of colour by putting something before the light which is like coloured glass; it is coloured gelatin. I will put in a plate of red gelatin which is carefully prepared of chemical materials and see what that will do. Of all the light passing to it from violet to red it only lets through the red and orange, giving a mixed reddish colour. Here is a plate of green gelatin : the green absorbs all the red, giving only green. Here is a plate absorbing something from each portion of the spectrum, taking away a great deal of the violet and giving a yellow or orange appearance to the light. Here is another absorbing the green and all the violet, leaving red, orange, and a very little faint green.

When the spectrum is very carefully produced, far more carefully than Newton knew how to show it, we have a homogeneous spectrum. It

Y 2

must be noticed that Newton did not understand what we call a homogeneous spectrum; he did not produce it, and does not point out in his writings the conditions for producing it. With an exceedingly fine line of light we can bring it out as in sunlight, like this upper picture—red, orange, yellow, green, blue, indigo, and violet, according to Newton's nomenclature. Newton never used a narrow beam of light, and so could not have had a homogeneous spectrum.

This is a diagram painted on glass and showing the colours as we know them. It would take two or three hours if I were to explain the subject of spectrum analysis to-night. We must tear ourselves away from it. I will just read out to you the wave-lengths corresponding to the different positions in the sun's spectrum of certain dark lines commonly called " Fraunhofer's lines." I will take as a unit the one hundred thousandth of a centimetre. A centimetre is 4 of an inch; it is a rather small half an inch. I take the thousandth of a centimetre and the hundredth of that as a unit. At the red end of the spectrum the light in the neighbourhood of that black line A (Fig. 48) has for its wave-length 7·6; B has 6·87; D has 5·89; the "frequency" for A is 3·9 times 100 million million, the frequency of D light is 5·1 times 100 million million per second.

Now what force is concerned in those vibrations as compared with sound at the rate of 400 vibrations per second? Suppose for a moment the same matter was to move to and fro through the same range but 400 million million times per second. The force required is as the square of the number expressing the frequency. Double frequency would require quadruple force for the vibration of the same body. Suppose I vibrate my hand again, as I did before. If I move it once per second a moderate force is required; for it to vibrate ten times per second 100 times as much force is required; for 400 vibrations per second 160,000 times as much force. If I move my hand once per second through a space of a quarter of an inch a very small force is required; it would require very considerable force to move it ten times a second, even through so small a range; but think of the force required to move a tuning-fork 400 times a second, and compare that with the force required for a motion of 400 million million times a second. If the mass moved is the same, and the range of motion is the same, then the force would be one million million million million times as great as the force required to move the prongs of the tuning-fork—it is as easy to understand that number as any number like 2, 3, or 4. Consider now what that number means and what we are to infer from it. What force is there in the space between my eye and that light? What forces are there in the space between our eyes and the sun, and our eyes and the remotest visible star? There is matter and there is motion, but what magnitude of force may there be?

I move through this "luminiferous ether" as if it were nothing. But were there vibrations with such frequency in a medium of steel or brass, they would be measured by millions and millions and millions of tons' action on a square inch of matter. There are no such forces in our air. Comets make a disturbance in the air, and perhaps the luminiferous ether is split up by the motion of a comet through it. So when we explain the nature of electricity, we explain it by a motion of the luminiferous ether. We cannot say that it is electricity. What can this luminiferous ether be? It is something that the planets move through with the greatest ease. It permeates our air; it is nearly in the same condition, so far as our means of judging are concerned, in our air and in the inter-planetary space. The air disturbs it but little; you may reduce air by air-pumps to the hundred thousandth of its density, and you make little effect in the transmission of light through it. The luminiferous ether is an elastic solid, for which the nearest analogy I can give you is this jelly which you see,[1] and the nearest analogy to the waves of light is the motion, which you can imagine, of this elastic jelly, with a ball of wood floating in the middle of it. Look there, when with my hand I vibrate the little red ball

[1] Exhibiting a large bowl of clear jelly with a small red wooden ball embedded in the surface near the centre.

up and down, or when I turn it quickly round the vertical diameter, alternately in opposite directions ;—that is the nearest representation I can give you of the vibrations of luminiferous ether.

Another illustration is Scottish shoemakers' wax or Burgundy pitch, but I know Scottish shoemakers' wax better. It is heavier than water, and absolutely answers my purpose. I take a large slab of the wax, place it in a glass jar filled with water, place a number of corks on the lower side and bullets on the upper side. It is brittle like the Trinidad pitch, or Burgundy pitch which I have in my hand—you can see how hard it is— but when left to itself it flows like a fluid. The shoemakers' wax breaks with a brittle fracture, but it is viscous and gradually yields.

What we know of the luminiferous ether is that it has the rigidity of a solid and gradually yields. Whether or not it is brittle and cracks we cannot yet tell, but I believe the discoveries in electricity and the motions of comets and the marvellous spurts of light from them, tend to show cracks in the luminiferous ether—show a correspondence between the electric flash and the aurora borealis and cracks in the luminiferous ether. Do not take this as an assertion, it is hardly more than a vague scientific dream: but you may regard the existence of the luminiferous ether as a reality of science; that is, we have an all-pervading medium, an elastic solid, with a great degree of rigidity—a rigidity so prodigious in proportion to its density that the vibrations of light in it have the frequencies I have mentioned, with the wave-lengths I have mentioned. The fundamental question as to whether or not luminiferous ether has gravity has not been answered. We have no knowledge that the luminiferous ether is attracted by gravity; it is sometimes called imponderable because some people vainly imagine that it has no weight: I call it matter with the same kind of rigidity that this elastic jelly has.

Here are two tourmalines; if you look through them toward the light you see the white light all round, *i.e.* they are transparent. If I turn round one of these tourmalines the light is extinguished, it is absolutely black, as though the tourmalines were opaque. This is an illustration of what is called polarisation of light. I cannot speak to you about qualities of light without speaking of the polarisation of light. I want to show you a most beautiful effect of polarising light, before illustrating a little further by means of this large mechanical illustration which you have in the bowl of jelly. What you saw first were two plates of the crystal tourmaline (which came from Brazil, I believe) having the property of letting light pass when both plates are placed in one particular direction as regards their axes of crystallisation, and extinguishing it when it passes through them with one of the plates held in another direction. Now I put in the lantern an instrument called a "Nicol prism," which also gives rays of polarised light. A Nicol prism is a piece of Iceland spar, cut in two and turned one part relatively to the other in a very ingenious way, and put together again and cemented into one by Canada balsam. The Nicol prism takes advantage of the property

which the spar has of double refraction, and
produces the phenomenon which I now show
you. I turn one prism round in a certain direction and you get light—a maximum of light. I
turn it through a right angle and you get blackness. I turn it one quarter round again, and
get maximum light ; one quarter more, maximum
blackness ; one quarter more, and bright light.
We rarely have such a grand specimen of a
Nicol prism as this.

There is another way of producing polarised
light. I stand before that light and look at its
reflection in a plate of glass on the table through
one of the Nicol prisms, which I turn round, so.
Now if I incline that plate of glass at a particular angle—rather more than fifty-five degrees—
I find a particular position in which, if I look at it
and then turn the prism round in the hand, the
effect is absolutely to extinguish the light in one
position of the prism and to give it maximum
brightness in another position. I use the term
"absolute" somewhat rashly. It is only a reduction to a very small quantity of light, not an

absolute annulment as we have in the case of the two Nicol prisms used conjointly. As to the mechanics of the thing, those of you who have never heard of this before would not know what I was talking about; it could only be explained to you by a course of lectures in physical optics. The thing is this, vibrations of light must be in a definite direction relatively to the line in which the light travels.

Look at this diagram, the light goes from left to right; we have vibrations perpendicular to the line of transmission. There is a line up and down which is the line of vibration. Imagine here a source of light, violet light, and here in front of it is the line of propagation. Sound-vibrations are to and fro in, this is transverse to, the line of propagation. Here is another, perpendicular to the diagram, still following the law of transverse vibration; here is another, circular vibration. Imagine a long rope, you whirl one end of it and you see a screw-like motion running along, and you can get this circular motion in one direction or in the opposite.

Plane-polarised light is light with the vibrations all in a single plane, perpendicular to the plane through the ray which is technically called the "plane of polarisation." Circularly polarised light consists of undulations of luminiferous ether having a circular motion. Elliptically polarised light is something between the two, not in a straight line, and not in a circular line; the course of vibration is an ellipse. Polarised light is light that performs its motions continually in one mode or direction. If in a straight line it is plane-polarised; if in a circular direction it is circularly polarised light; when elliptical it is elliptically polarised light.

With Iceland spar, one unpolarised ray of light divides on entering it into two rays of polarised light, by reason of its power of double refraction, and the vibrations are perpendicular to one another in the two emerging rays. Light is always polarised when it is reflected from a plate of unsilvered glass, or from water, at a certain definite angle of fifty-six degrees for glass, fifty-two degrees for water, the angle being reckoned in each case from a perpendicular to the surface. The angle for water is the angle whose tangent is 1·4. I wish you to look at the polarisation with your own eyes. Light from glass at fifty-six degrees and from water at fifty-two degrees goes away vibrating perpendicularly to the plane of incidence and plane of reflection.

We can distinguish it without the aid of an instrument. There is a phenomenon well known in physical optics as "Haidinger's Brushes." The discoverer is well known in Philadelphia as a mineralogist, and the phenomenon I speak of goes by his name. Look at the sky in a direction of ninety degrees from the sun, and you will see a yellow and blue cross, with the yellow toward the sun, and from the sun, spreading out like two foxes' tails with blue between, and then two red brushes in the space at right angles to the blue. If you do not see it, it is because your eyes are not sensitive enough, but a little training will give them the needed sensitiveness. If you cannot see it in this way try another method. Look into a pail of water with a black bottom; or take a clear glass dish of water, rest it on a black cloth, and look down at the surface of the water on a day with a white cloudy sky (if there is such a thing ever to be seen in Philadelphia). You will see the white sky reflected in the basin of water at an angle of about fifty degrees. Look at it with the head tipped on one side and then again with the head tipped to the other side, keeping your eyes on the water, and you will see Haidinger's brushes. Do not do it fast or you will make yourself giddy. The explanation of this is the refreshing of the sensibility of the retina. The Haidinger's brush is always there, but you do not see it because your eye is not sensitive enough. After once seeing it you always see it ; it does not thrust itself inconveniently before you when you do not want to see it. You can also readily see it in a piece of glass with a dark cloth below it, or in a basin of water.

I am going to conclude by telling you how we know the wave-lengths of light, and how we know the frequency of the vibrations, and we shall actually make a measurement of the wave-length of yellow light. I am now going to show you the diffraction spectrum.

You see on the screen,[1] on each side of a central white bar of light, a set of bars of light of variegated colours, the first one on each side showing blue or indigo colour, about four inches from the central white bar, and red about four inches farther, with vivid green between the blue and the red. That effect is produced by a grating with 400 lines to the centimetre, engraved on glass, which I now hold in my hand. The next grating that we shall try has 3,000 lines on a Paris inch. You see the central space and on each side a large number of spectrums, blue at one end and red at the other. The fact that, in the first spectrum, red is about twice as far from the centre as the blue, proves that a wave-length of red light is double that of blue light.

I will now show you the operation of measuring the length of a wave of sodium light, that is a light like that marked D on the spectrum

[1] Showing the chromatic bands thrown upon the screen from a diffraction grating.

(Fig. 48), a light produced by a spirit-lamp with
salt in it. The sodium vapour is heated up to
several thousand degrees, when it becomes self-
luminous and gives such a light as we get by
throwing salt upon a spirit lamp in the game of
snap-dragon.

I hold in my hand a beautiful grating of glass
silvered by Liebig's process with metallic silver,
a grating with 6,480 lines to the inch, belonging to
my friend Professor Barker, which he has kindly
brought here for us this evening. You will see
the brilliancy of colour as I turn the light reflected
from the grating toward you and pass the beam
round the room. You have now seen directly with
your own eyes these brilliant colours reflected from
the grating, and you have also seen them thrown
upon the screen from a grating placed in the
lantern. Now with a grating of 17;000 lines per
inch—a much greater number than the other—
you will see how much further from the central
bright space the first spectrum is ; how much more
this grating changes the direction, or diffraction,
of the beam of light. Here is the centre of the

z

grating, and there is the first spectrum. You will note that the violet light is least diffracted and the red light is most diffracted. This diffraction of light first proved to us definitely the reality of the undulatory theory of light.

You ask why does not light go round a corner as sound does. Light does go round a corner in these diffraction spectrums; and it is shown going round a corner, since it passes through these bars and is turned round an angle of thirty degrees. The phenomena of light going round a corner seen by means of instruments adapted to show the result and to measure the angles through which it is turned, is called the diffraction of light.

I can show you an instrument which will measure the wave-lengths of light. Without proving the formula, let me tell it to you. A spirit-lamp with salt sprinkled on the wick gives very nearly homogeneous light, that is to say, light of one wave length, or all of the same period. I have here a little grating which I take in my hand. I look through this grating and see that candle before me. Close behind it you see a blackened slip of wood with two white marks on it ten inches asunder. The line on which they are marked is placed perpendicular to the line at which I shall go from it. When I look at this salted spirit-lamp I see a series of spectrums of yellow light. As I am somewhat short-sighted I am making my eye see with this eye-glass and the natural lenses of the eye what a long-sighted person would make out without an eye-glass. On that screen you saw a succession of spectrums. I now look direct at the candle and what do I see? I see a succession of five or six brilliantly coloured spectrums on each side of the candle. But when I look at the salted spirit-lamp, now I see ten spectrums on one side and ten on the other, each of which is a monochromatic band of light.

I will measure the wave length of the light thus. I walk away to a considerable distance and look at the spirit-lamp and marks. I see a set of spectrums. The first white line is exactly behind the flame. I want the first spectrum to the right of that white line to fall exactly on the other white line, which is ten inches from the first. As I walk away

Z 2

from it I see it is now very near it ; it is now on it.
Now the distance from my eye is to be measured,
and the problem is again to reduce feet to inches.
The distance from the spectrum of the flame to
my eye is thirty-four feet nine inches. Mr.
President, how many inches is that ? 417 inches,
in round numbers 420 inches. Then we have the
proportion, as 420 is to 10 so is the length from
bar to bar of the grating to the wave-length
of sodium light. That is to say as forty-two is
to one. The distance from bar to bar is the four
hundredth of a centimetre : therefore the 42nd
part of the four hundredth of a centimetre is the
wave-length according to our simple, and easy
and hasty experiment. The true wave-length
of sodium light, according to the most accurate
measurement, is about a 17,000th of a centimetre,
which differs by scarcely more than one per cent
from our result !

The only apparatus you see is this little grating
—a piece of glass having a space four-tenths of
an inch wide ruled with 400 fine lines. Any of
you who will take the trouble to buy one may

measure the wave-length of a candle flame himself. I hope some of you will be induced to make the experiment for yourselves.

If I put salt on the flame of a spirit lamp, what do I see through this grating? I see merely a sharply defined yellow light, constituting the spectrum of vaporised sodium, while from the candle flame I see an exquisitely coloured spectrum, far more beautiful than that I showed you on the screen. I see in fact a series of spectrums on the two sides with the blue toward the candle flame and the red further out. I cannot get one definite thing to measure from in the spectrum from the candle flame, as I can with the flame of a spirit lamp with the salt thrown on it, which gives as I have said a simple yellow light. The highest blue light I see in the candle flame is now exactly on the line. Now measure to my eye, it is forty-four feet four inches, or 532 inches. The length of this wave then is the 532d part of the four hundredth of a centimetre which would be the 21,280th of a centimetre, say the 21,000th of a centimetre. Then measure for the red and you will find something like the 11,000th for the lowest of the red light.

Lastly, how do we know the frequency of vibration?

Why, by the velocity of light. How do we know that? We know it in a number of different ways, which I cannot explain now because time forbids, and I can now only tell you shortly that the frequency of vibration for any particular ray is equal to the velocity of light divided by the wave-length for that ray. The velocity of light is about 187,000 British statute miles per second, but it is much better to take the kilometre— which is about six-tenths of a mile—for the unit, when we find the velocity is very accurately 300,000 kilometres, or 30,000,000,000 centimetres, per second. Take now the wave-length of sodium light, as we have just measured it by means of the salted spirit lamp, to be one 17,000th of a centimetre, and we find the frequency of vibration of the sodium light to be 510 million million per second. There, then, you have a calculation of the frequency from a simple observation which you all can make for yourselves.

Lastly, I must tell you about the colour of the blue sky which is illustrated by this spherule imbedded in an elastic solid (Fig. 49). I want to explain to you in two minutes the mode of vibration. Take the simplest plane-polarised light. Here is a spherule which is producing it in an

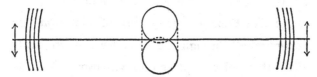

FIG. 49.—Vibrating Spherule Imbedded in an Elastic Solid.

elastic solid. Imagine the solid to extend miles horizontally and miles up and down, and imagine this spherule to vibrate up and down. It is quite clear that it will make transverse vibrations similarly in all horizontal directions. The plane of polarisation is defined as a plane perpendicular to the line of vibration. Thus, light produced by a molecule vibrating up and down, as this red globe in the jelly before you, is polarised in a horizontal plane because the vibrations are vertical.

Here is another mode of vibration. Let me twist this spherule in the jelly as I am now doing, and that will produce vibrations, also spreading out equally in all horizontal directions. When I twist this globe round it draws the jelly round with it; twist it rapidly back and the jelly flies back. By the inertia of the jelly the vibrations spread in all directions and the lines of vibration are horizontal all through the jelly. Everywhere, miles away that solid is placed in vibration. You do not see the vibrations, but you must understand that they are there. If it flies back it makes vibration, and we have waves of horizontal vibrations travelling out in all directions from the exciting molecule.

I am now causing the red globe to vibrate to and fro horizontally. That will cause vibrations to be produced which will be parallel to the line of motion at all places of the plane perpendicular to the range of the exciting molecule. What makes the blue sky? These are exactly the motions that make the blue light of the sky, which is due to spherules in the luminiferous ether, but little modified by the air. Think of the sun near the horizon, think of the light of the sun streaming through and giving you the azure blue and violet overhead. Think first of any one particle and think of it moving in such a way as to give horizontal and vertical vibrations and circular and elliptic vibrations.

You see the blue sky in high pressure steam blown into the air; you see it in the experiment of Tyndall's blue sky in which a delicate condensation of vapour gives rise to exactly the azure blue of the sky.

Now the motion of the luminiferous ether relatively to the spherule gives rise to the same effect as would an opposite motion impressed upon the spherule quite independently by an independent force. So you may think of the blue colour coming from the sky as being produced by to and fro vibrations of matter in the air, which vibrates much as this little globe vibrates imbedded in the jelly.

The result in a general way is this: The light coming from the blue sky is polarised in a plane through the sun, but the blue light of the sky is complicated by a great number of circumstances and one of them is this, that the air is illuminated not only by the sun but by the earth. If we could get the earth covered by a black cloth then we could study the polarised light of the sky with a simplicity which we cannot do now. There are, in nature, reflections from seas and rocks and hills and waters in an infinitely complicated manner.

Let observers observe the blue sky not only in winter when the earth is covered with snow, but in summer when it is covered with dark green foliage. This will help to unravel the complicated phenomena in question. But the azure blue of the sky is light produced by the reaction on the vibrating ether of little spherules of water, of perhaps a fifty thousandth or a hundred thousandth of a centimetre diameter, or perhaps little motes, or lumps, or crystals of common salt, or particles of dust, or germs of vegetable or animal species wafted about in the air. Now what is the luminiferous ether ? It is matter prodigiously less dense than air—millions and millions and millions of times less dense than air. We can form some sort of idea of its limitations. We believe it is a real thing, with great rigidity in comparison with its density : it may be made to vibrate 400 million million times per second ; and yet be of such density as not to produce the slightest resistance to any body going through it.

Going back to the illustration of the shoemaker's wax ; if a cork will, in the course of a year, push its way up through a plate of that wax when placed under water, and if a lead bullet will penetrate downwards to the bottom, what is the law of the resistance ? It clearly depends on time. The cork slowly in the course of a year works its way up through two inches of that substance; give it one or two thousand years to do it and the resistance will be enormously less ; thus the motion of a cork or bullet, at the rate of one inch in 2,000 years, may be compared with that of the earth, moving at the rate of six times ninety-three million miles a year, or nineteen miles per second,

through the luminiferous ether ; but when we can
have actually before us a thing elastic like jelly
and yielding like pitch, surely we have a large
and solid ground for our faith in the speculative
hypothesis of an elastic luminiferous ether, which
constitutes the wave theory of light.

ON THE AGE OF THE SUN'S HEAT.

[*Reprinted by permission from "Macmillan's Magazine,"* *March*, 1862.]

THE second great law of Thermodynamics involves a certain principle of *irreversible action in nature*. It is thus shown that, although mechanical energy is *indestructible*, there is a universal tendency to its dissipation, which produces gradual augmentation and diffusion of heat, cessation of motion, and exhaustion of potential energy through the material universe.[1] The result would inevitably be a state of universal rest and death, if the universe were finite and left to obey existing laws. But it is impossible to conceive a limit to the extent of matter in the

[1] See "On a Universal Tendency in Nature to the Dissipation of Mechanical Energy," Proceedings of the Royal Society of Edinburgh, April 19, 1852; or the *Philosophical Magazine*, October, 1852; also *Mathematical and Physical Papers*, Vol. I. Article LIX.

universe ; and therefore science points rather to an endless progress, through an endless space, of action involving the transformation of potential energy into palpable motion and thence into heat, than to a single finite mechanism, running down like a clock, and stopping for ever. It is also impossible to conceive either the beginning or the continuance of life, without an overruling creative power ; and, therefore, no conclusions of dynamical science regarding the future condition of the earth can be held to give dispiriting views as to the destiny of the race of intelligent beings by which it is at present inhabited.

The object proposed in the present article is an application of these general principles to the discovery of probable limits to the periods of time, past and future, during which the sun can be reckoned on as a source of heat and light. The subject will be discussed under three heads :—

I. The secular cooling of the sun.

II. The present temperature of the sun.

III. The origin and total amount of the sun's heat.

PART I.

ON THE SECULAR COOLING OF THE SUN.

How much the sun is actually cooled from year to year, if at all, we have no means of ascertaining, or scarcely even of estimating in the roughest manner. In the first place we do not know that he is losing heat at all. For it is quite certain that *some heat* is generated in his atmosphere by the influx of meteoric matter; and it is possible that the *amount* of heat so generated from year to year is sufficient to compensate the loss by radiation. It is, however, also possible that the sun is now an incandescent liquid mass, radiating away heat, either primitively created in his substance, or, what seems far more probable, generated by the falling in of meteors in past times, with no sensible compensation by a continuance of meteoric action.

It has been shown [1] that, if the former sup-

---

[1] "On the Mechanical Energies of the Solar System," Transactions of the Royal Society of Edinburgh, April, 1854, and

position were true, the meteors by which the sun's heat would have been produced during the last 2,000 or 3,000 years must have been all that time much within the earth's distance from the sun, and must therefore have approached the central body in very gradual spirals; because, if enough of matter to produce the supposed thermal effect fell in from space outside the earth's orbit, the length of the year would have been very sensibly shortened by the additions to the sun's mass which must have been made. The quantity of matter annually falling in must, on that supposition, have amounted to $1/47$ of the earth's mass, or to $1/15,000,000$ of the sun's; and therefore it would be necessary to suppose the "Zodiacal Light" to amount to at least $1/5,000$ of the sun's mass, to account in the same way for a future supply of 3,000 years' sun-heat. When these conclusions were first published it was pointed out that "disturbances in the motions of visible planets" should be looked for, as affording us means for

Philosophical Magazine, December, 1854 (*Mathematical and Physical Papers,* Vol. II., Article LXVI.).

estimating the possible amount of matter in the zodiacal light; and it was conjectured that it could not be nearly enough to give a supply of 30,000 years' heat at the present rate. These anticipations have been to some extent fulfilled in Le Verrier's great researches on the motion of the planet Mercury, which have recently given evidence of a sensible influence attributable to matter circulating, as a great number of small planets, within his orbit round the sun. But the amount of matter thus indicated is very small; and, therefore, if the meteoric influx taking place at present is enough to produce any appreciable portion of the heat radiated away, it must be supposed to come from matter circulating round the sun, within very short distances of his surface. The density of this meteoric cloud would have to be supposed so great that comets could scarcely have escaped as comets actually have escaped, showing no discoverable effects of resistance, after passing his surface within a distance equal to 1/8 of his radius. All things considered, there seems little probability in the hypothesis that

A A

solar radiation is at present compensated, to any appreciable degree, by heat generated by meteors falling in ; and, as it can be shown that no chemical theory is tenable,[1] it must be concluded as most probable that the sun is at present merely an incandescent liquid mass cooling.

How much he cools from year to year, becomes therefore a question of very serious import, but it is one which we are at present quite unable to answer. It is true we have data on which we might plausibly found a probable estimate, and from which we might deduce, with at first sight seemingly well-founded confidence, limits, not very wide, within which the present true rate of the sun's cooling must lie. For we know, from the independent but concordant investigations of Herschel and Pouillet, that the sun radiates every year from his whole surface about 6×10^{30} (six million million million million million) times as much heat as is sufficient to raise the temperature of 1 lb. of water by 1° Cent. We also have excellent reason for believing that the sun's

[1] " Mechanical Energies of the Solar System." See note p. 351.

substance is very much like the earth's. Stokes's principles of solar and stellar chemistry have been for many years explained in the University of Glasgow, and it has been taught as a first result that sodium does certainly exist in the sun's atmosphere, and in the atmospheres of many of the stars, but that it is not discoverable in others. The recent application of these principles in the splendid researches of Bunsen and Kirchhof (who made an independent discovery of Stokes's theory) has demonstrated with equal certainty that there are iron and manganese, and several of our other known metals, in the sun. The specific heat of each of these substances is less than the specific heat of water, which indeed exceeds that of every other known terrestrial body, solid or liquid. It might, therefore, at first sight seem probable that the mean specific heat [1] of the sun's whole sub-

[1] The "specific heat" of a homogeneous body is the quantity of heat that a unit of its substance must acquire or must part with, to rise or to fall by 1° in temperature. The mean specific heat of a heterogeneous mass, or of a mass of homogeneous substance, under different pressures in different parts, is the quantity of heat which the whole body takes or gives in rising or in falling 1° in temperature, divided by the number of units in its mass. The expression,

stance is less, and very certain that it cannot be much greater, than that of water. If it were equal to the specific heat of water we should only have to divide the preceding number (6×10^{30}), derived from Herschel's and Pouillet's observations, by the number of pounds (4.3×10^{30}) in the sun's mass, to find $1°.4$ Cent. for the present annual rate of cooling. It might therefore seem probable that the sun cools more, and almost certain that he does not cool less, than a centigrade degree and four-tenths annually. But, if this estimate were well-founded, it would be equally just to assume that the sun's expansibility[1] with heat does not differ greatly from that of some average terrestrial

"mean specific heat" of the sun, in the text, signifies the total amount of heat actually radiated away from the sun, divided by his mass, during any time in which the average temperature of his mass sinks by $1°$, whatever physical or chemical changes any part of his substance may experience.

[1] The "expansibility in volume," or the "cubical expansibility," of a body, is an expression technically used to denote the proportion which the increase or diminution of its bulk, accompanying a rise or fall of $1°$ in its temperature, bears to its whole bulk at some stated temperature. The expression, "the sun's expansibility," used in the text, may be taken as signifying the ratio which the actual contraction, during a lowering of his mean temperature by $1°$ Cent., bears to his present volume.

body. If, for instance, it were the same as that of
solid glass, which is about 1/40,000 on bulk, or
1/120,000 on diameter, per 1° Cent. (and for most
terrestrial liquids, especially at high temperatures,
the expansibility is much more), and if the specific
heat were the same as that of liquid water, there
would be in 860 years a contraction of 1 per cent.
on the sun's diameter, which could scarely have
escaped detection by astronomical observation.
There is, however, a far stronger reason than this
for believing that no such amount of contraction
could have taken place, and therefore for suspect-
ing that the physical circumstances of the sun's
mass render the condition of the substances of
which it is composed, as to expansibility and
specific heat, very different from that of the same
substances when experimented on in our terrestrial
laboratories. Mutual gravitation between the
different parts of the sun's contracting mass
must do an amount of work, which cannot
be calculated with certainty, only because the
law of the sun's interior density is not known.
The amount of work performed on a contrac-

tion of one-tenth per cent. of the diameter, if the density remained uniform throughout the interior, would, as Helmholtz showed, be equal to 20,000 times the mechanical equivalent of the amount of heat which Pouillet estimated to be radiated from the sun in a year. But in reality the sun's density must increase very much towards his centre, and probably in varying proportions, as the temperature becomes lower and the whole mass contracts. We cannot, therefore, say whether the work actually done by mutual gravitation during a contraction of one-tenth per cent. of the diameter, would be more or less than the equivalent of 20,000 years' heat ; but we may regard it as most probably not many times more or less than this amount. Now, it is in the highest degree improbable that mechanical energy can in any case increase in a body contracting in virtue of cooling. It is certain that it really does diminish very notably in every case hitherto experimented on. It must be supposed, therefore, that the sun always radiates away in heat something more than the Joule-equivalent of the work done on his contracting mass, by mutual gravitation of its parts. Hence, in contracting by one-tenth per cent. in his diameter, or three-tenths per cent. in his bulk, the sun must give out something either more, or not greatly less, than 20,000 years' heat ; and thus, even without historical evidence as to the constancy of his diameter, it seems safe to conclude that no such contraction as that calculated above (one per cent. in 860 years), can have taken place in reality. It seems, on the contrary, probable that, at the present rate of radiation, a contraction of one-tenth per cent. in the sun's diameter could not take place in much less than 20,000 years, and scarcely possible that it could take place in less than 8,600 years. If then, the mean specific heat of the sun's mass, in its actual condition, is not more than ten times that of water, the expansibility in volume must be less than $1/4000$ per $100°$ Cent., (that is to say, less than $1/10$ of that of solid glass,) which seems improbable. But although from this consideration we are led to regard it as probable that the sun's specific heat is considerably more than ten times that of water (and, therefore, that his mass cools considerably less than 100° C. in 700 years, a conclusion which, indeed, we could scarcely avoid on simply geological grounds), the physical principles we now rest on fail to give us any reason for supposing that the sun's specific heat is more than 10,000 times that of water, because we cannot say that his expansibility in volume is probably more than 1/400 per 1° Cent. And there is, on other grounds, very strong reason for believing that the specific heat is really much less than 10,000. For it is almost certain that the sun's mean temperature is even now as high as 14,000° Cent. ; and the greatest quantity of heat that we can explain, with any probability, to have been by natural causes ever acquired by the sun (as we shall see in the third part of this article), could not have raised his mass at any time to this temperature, unless his specific heat were less than 10,000 times that of water.

We may therefore consider it as rendered highly probable that the sun's specific heat is more than ten times, and less than 10,000 times, that of liquid water. From this it would follow with certainty that his temperature sinks 100° Cent. in some time from 700 years to 700,000 years.

What then are we to think of such geological estimates as 300,000,000 years for the "denudation of the Weald"? Whether is it more probable that the physical conditions of the sun's matter differ 1,000 times more than dynamics compel us to suppose they differ from those of matter in our laboratories; or that a stormy sea, with possibly channel tides of extreme violence, should encroach on a chalk cliff 1,000 times more rapidly than Mr. Darwin's estimate of one inch per century?

Part II.

ON THE SUN'S PRESENT TEMPERATURE.

At his surface the sun's temperature cannot, as we have many reasons for believing, be incomparably higher than temperatures attainable artifically in our terrestrial laboratories.

Among other reasons it may be mentioned that

the sun radiates out heat from every square foot of his surface, at only about 7,000 horse power.[1] Coal, burning at a rate of a little less than a pound per two seconds, would generate the same amount; and it is estimated (Rankine, *Prime Movers*, p. 285, Ed. 1852) that, in the furnaces of locomotive engines, coal burns at from one pound in thirty seconds to one pound in ninety seconds, per square foot of grate-bars. Hence heat is radiated from the sun at a rate not more than from fifteen to forty-five times as high as that at which heat is generated on the grate-bars of a locomotive furnace, per equal areas.

The interior temperature of the sun is probably far higher than that at his surface, because direct conduction can play no sensible part in the transference of heat between the inner and outer portions of his mass, and there must be an

---

[1] One horse power in mechanics is a technical expression (following Watt's estimate), used to denote a rate of working in which energy is involved at the rate of 33,000 foot pounds per minute. This, according to Joule's determination of the dynamical value of heat, would, if spent wholly in heat, be sufficient to raise the temperature of 23¾ lbs. of water by 1° Cent. per minute.

approximate *convective* equilibrium of heat throughout the whole, if the whole is fluid. That is to
say, the temperatures, at different distances from
the centre, must be approximately those which
any portion of the substance, if carried from the
centre to the surface, would acquire by expansion
without loss or gain of heat.

PART III.

ON THE ORIGIN AND TOTAL AMOUNT OF THE SUN'S HEAT.

The sun being, for reasons referred to above,
assumed to be an incandescent liquid now losing
heat, the question naturally occurs, How did this
heat originate? It is certain that it cannot have
existed in the sun through an infinity of past
time, since, as long as it has so existed, it must
have been suffering dissipation, and the finiteness
of the sun precludes the supposition of an infinite
primitive store of heat in his body.

The sun must, therefore, either have been

created an as active source of heat at some time
of not immeasurable antiquity, by an over-ruling
decree ; or the heat which he has already radiated
away, and that which he still possesses, must
have been acquired by a natural process, following permanently established laws. Without pro-
nouncing the former supposition to be essentially
incredible, we may safely say that it is in the
highest degree improbable, if we can show
the latter to be not contradictory to known
physical laws. And we do show this and more,
by merely pointing to certain actions, going on
before us at present, which, if sufficiently abundant
at some past time, must have given the sun heat
enough to account for all we know of his past
radiation and present temperature.

It is not necessary at present to enter at length
on details regarding the meteoric theory, which
appears to have been first proposed in a definite
form by Mayer, and afterwards independently
by Waterston ; or regarding the modified hypothesis of meteoric vortices, which the writer of the
present article showed to be necessary, in order that

the length of the year, as known for the last 2,000 years, may not have been sensibly disturbed by the accessions which the sun's mass must have had during that period, if the heat radiated away has been always compensated by heat generated by meteoric influx.

For reasons mentioned in the first part of the present article, we may now believe that all theories of complete, or nearly complete, contemporaneous meteoric compensation, must be rejected ; but we may still hold that—

"Meteoric action . . . . is . . . . not only proved
" to exist as a cause of solar heat, but it is the only
" one of all conceivable causes which we know to
" exist from independent evidence." [1]

The form of meteoric theory which now seems most probable, and which was first discussed on true thermodynamic principles by Helmholtz,[2] consists in supposing the sun and his heat to have originated in a coalition of smaller bodies, falling together by mutual gravitation, and

[1] "Mechanical Energies of the Solar System." See note p. 351.
[2] Popular lecture delivered on the 7th February, 1854, at Königsberg, on the occasion of the Kant commemoration.

generating, as they must do according to the great law demonstrated by Joule, an exact equivalent of heat for the motion lost in collision.

That some form of the meteoric theory is certainly the true and complete explanation of solar heat can scarely be doubted, when the following reasons are considered :

(1). No other natural explanation, except by chemical action, can be conceived.

(2). The chemical theory is quite insufficient, because the most energetic chemical action we know, taking place between substances amounting to the whole sun's mass, would only generate about 3,000 years' heat.[1]

(3). There is no difficulty in accounting for 20,000,000 years' heat by the meteoric theory.

It would extend this article to too great a length, and would require something of mathematical calculation, to explain fully the principles on which this last estimate is founded. It is enough to say that bodies, all much smaller than the sun, falling together from a state of relative rest,

[1] " Mechanical Energies of the Solar System." See note p. 351.

at mutual distances all large in comparison with
their diameters, and forming a globe of uniform
density equal in mass and diameter to the sun,
would generate an amount of heat which, accurately calculated according to Joule's principles
and experimental results, is found to be just
20,000,000 times Pouillet's estimate of the annual
amount of solar radiation. The sun's density
must, in all probability, increase very much towards
his centre, and therefore a considerably greater
amount of heat than that must be supposed to
have been generated if his whole mass was
formed by the coalition of comparatively small
bodies. On the other hand, we do not know
how much heat may have been dissipated by resistance and minor impacts before the final
conglomeration; but there is reason to believe
that even the most rapid conglomeration that we
can conceive to have probably taken place, could
only leave the finished globe with about half the
entire heat due to the amount of potential energy
of mutual gravitation exhausted. We may, therefore, accept, as a lowest estimate for the sun's

initial heat, 10,000,000 times a year's supply at present rate, but 50,000,000 or 100,000,000 as possible, in consequence of the sun's greater density in his central parts.

The considerations adduced above, in this paper, regarding the sun's possible specific heat, rate of cooling, and superficial temperature, render it probable that he must have been very sensibly warmer one million years ago than now; and, consequently, if he has existed as a luminary for ten or twenty million years, he must have radiated away considerably more than the corresponding number of times the present yearly amount of loss.

It seems, therefore, on the whole most probable that the sun has not illuminated the earth for 100,000,000 years, and almost certain that he has not done so for 500,000,000 years. As for the future, we may say, with equal certainty, that inhabitants of the earth cannot continue to enjoy the light and heat essential to their life, for many million years longer, unless sources now unknown to us are prepared in the great storehouse of creation.

ON THE SUN'S HEAT.

A Friday evening Lecture delivered before the Royal Institution of Great Britain on January 21, 1887 : *see also* Good Words *for March and April* 1887.]

FROM human history we know that for several thousand years the sun has been giving heat and light to the earth as at present, possibly with some considerable fluctuations, and possibly with some not very small progressive variation. The records of agriculture, and the natural history of plants and animals within the time of human history, abound with evidence that there has been no exceedingly great change in the intensity of the sun's heat and light within the last three thousand years; but for all that, there may have been variations of quite as much as 5 or 10 per cent., as we may judge by considering that the intensity of the solar radiation to the earth is $6\frac{1}{2}$ per cent.

B B

greater in January than in July; and neither at the equator nor in the northern or southern hemispheres has this difference been discovered by experience or general observation of any kind. But as for the mere age of the sun, irrespective of the question of uniformity, we have proof of something vastly more than three thousand years in geological history, with its irrefragable evidence of continuity of life on the earth in time past for tens of thousands, and probably for millions of years.

Here, then, we have a splendid subject for contemplation and research in Natural Philosophy or Physics—the science of dead matter. The sun, a mere piece of matter of the moderate dimensions which we know it to have, bounded all round by cold ether,[1] has been doing work at the rate of four

---

[1] The sun warms and lights the earth by wave motion, excited in virtue of his white-hot temperature, and transmitted through a material commonly called the luminiferous ether, which fills all space as far as the remotest star, and has the property of transmitting radiant heat (or light) without itself becoming heated. I feel that I have a right to drop the adjective luminiferous, because the medium, far above the earth's surface, through which we receive sun-heat (or light), and through which the planets move, was called

hundred and seventy-six thousand million million million horse-power for three thousand years, and at possibly a higher, certainly not much lower, rate for a few million years. How is this to be explained? Natural philosophy cannot evade the question, and no physicist who is not engaged in trying to answer it, can have any other justification than that his whole working time is occupied with work on some other subject or subjects of his province, by which he has more hope of being able to advance science.

It may be taken as an established result of scientific inquiry that the sun is *not* a burning fire, and *is* merely a white-hot fluid mass cooling, with some little accession of fresh energy by meteors occasionally falling in, but of very small account in comparison with the whole energy of heat which he gives out from year to year. Helmholtz's form

ether 2,000 years before chemists usurped the name for "sulphuric ether," "muriatic ether," and other compounds, fancifully supposed to be peculiarly ethereal; and I trust that chemists of the present day will not be angry with me if I use the word ether pure and simple, to denote the medium whose undulatory motions constitute radiant heat (or light).

of the meteoric theory of the origin of the sun's heat, may be accepted as having the highest degree of scientific probability that can be assigned to any assumption regarding actions of prehistoric times. The essential principle of the explanation is this : at some period of time, long past, the sun's initial heat was generated by the collision of pieces of matter gravitationally attracted together from distant space to build up his present mass ; and shrinkage due to cooling gives, through the work done by the mutual gravitation of all parts of the shrinking mass, the vast heat-storage capacity in virtue of which the cooling has been, and continues to be, so slow.

In some otherwise excellent books it is "paradoxically" stated that the sun is becoming hotter because of the condensation.[1] Paradoxes

---

[1] [Note of February 21, 1887.—The "paradox" referred to here, is, as I now find, merely a misstatement (faulty and manifestly paradoxical through the omission of an essential condition) of an astonishing and most important conclusion of a paper by J. Homer Lane, which appeared in the *American Journal of Science*, for July, 1870 (referred to more particularly on p. 398 below). In Newcomb's *Popular Astronomy*, first edition, p. 508, the omission is supplied in a footnote, giving a clear popular explanation of the dynamics of

have no place in science. Their removal is the substitution of true for false statements and thoughts, not always so easily effected as in the present case. The truth is, that it is because the sun is becoming less hot *in places of equal density*, that his mass is allowed to yield gradually under the condensing tendency of gravity ; and thus from age to age cooling and condensation go on together.

An essential detail of Helmholtz's theory of solar heat is that the sun must be fluid, because even though at any given moment hot enough from the surface to any depth, however great, inwards, to be brilliantly incandescent, the conduction of heat from within through solid matter of even the highest conducting quality known to us, would not suffice to maintain the incandescence of the surface for more than a

Lane's conclusion ; and the subject is similarly explained in Ball's *Story of the Heavens*, pp. 501, 502, and 503, with complete avoidance of the "paradox." And now I take this opportunity of correcting my hasty correction of the "paradox" by the insertion of the five words in italics added to lines 8 and 9 of the paragraph. —W. T.]

few hours, after which all would be darkness. Observation confirms this conclusion so far as the outward appearance of the sun is concerned, but does not suffice to disprove the idea which was so eloquently set forth by Sir John Herschel, and which prevailed till thirty or forty years ago, that the sun is a solid nucleus inclosed in a sheet of violently agitated flame. In reality, the matter of the outer shell of the sun, from which the heat is radiated outwards, must in cooling become denser, and so becoming unstable in its high position must fall down, and hotter fluid from within must rush up to take its place. The tremendous currents thus continually produced in this great mass of flaming fluid constitute the province of the newly-developed science of solar physics, which, with its marvellous instrument of research —the spectroscope— is yearly and daily giving us more and more knowledge of the actual motions of the different ingredients, and of the splendid and all-important resulting phenomena.

To form some idea of the amount of the heat

which is being continually carried up to the sun's surface and radiated out into space, and of the dynamical relations between it and the solar gravitation, let us first divide that prodigious number (476×10^{21}) of horse-power by the number ($6\cdot1 \times 10^{18}$) of square metres[1] in the sun's surface, and we find 78,000 horse-power

---

[1] A square metre is about $10\frac{3}{4}$ (more nearly $10\cdot764$) square feet, or a square yard and a fifth (more nearly 1.196 square yards). The metre is a little less than 40 inches (39.37 inches $= 3.281$ feet $= 1.094$ yards). The kilometre, which we shall have to use presently, being a thousand metres, is a short mile as it were ($\cdot6214$ of the British statute mile). Thus in round numbers 62 statute miles is equal to 100 kilometres, and 161 kilometres is equal to 100 statute miles. The awful and unnecessary toil and waste of brain power involved in the use of the British system of inches, feet, yards, perches or rods or poles, chains, furlongs, British statute miles, nautical miles, square rod ($30\frac{1}{4}$ square yards)! rood (1210 square yards)! acre (4 roods), may be my apology, but it is only a part of my reason, for not reckoning the sun's area in acres, his activity in horse-power per square inch or per square foot, and his radius, and the earth's distance from him in British statute miles, and for using exclusively the one-denominational system introduced by the French ninety years ago, and now in common use in every civilised country of the world, except England and the United States of North America. The British ton is 1.016 times the French ton, or weight of a cubic metre of cold water (1016 kilogrammes). The French ton, of 1000 kilogrammes, is $\cdot9842$ of the British ton. Thus for many practical reckonings, such as those of the present paper, the difference between the British and the French ton may be neglected.

as the mechanical value of the radiation per square metre. Imagine, then, the engines of eight ironclads applied, by ideal mechanism of countless shafts, pulleys, and belts, to do all their available work of, say 10,000 horse-power each, in perpetuity driving one small paddle in a fluid contained in a square-metre vat. The same heat would be given out from the square-metre surface of the fluid as is given out from every square metre of the sun's surface.

But now to pass from a practically impossible combination of engines, and a physically impossible paddle and fluid and containing vessel, towards a more practical combination of matter for producing the same effect: still keep the ideal vat and paddle and fluid, but place the vat on the surface of a cool, solid, homogeneous globe of the same size (697,000 kilometres radius) as the sun, and of density (1.4) equal to the sun's mean density. Instead of using steam-power, let the paddle be driven by a weight descending in a pit excavated below the vat. As the simplest possible mechanism, take a long vertical shaft, with the paddle mounted on the top of it so as to turn horizontally. Let the weight be a nut working on a screw-thread on the vertical shaft, with guides to prevent the nut from turning—the screw and the guides being all absolutely frictionless. Let the pit be a metre square at its upper end, and let it be excavated quite down to the sun's centre, everywhere of square horizontal section, and tapering uniformly to a point in the centre. Let the weight be simply the excavated matter of the sun's mass, with merely a little clearance space between it and the four sides of the pit, and with a kilometre or so cut off the lower pointed end to allow space for its descent. The mass of this weight is 326 million tons. Its heaviness, three-quarters of the heaviness of an equal mass at the sun's surface, is 244 million tons solar surface-heaviness. Now a horse-power is, per hour, 270 metre-tons, terrestrial surface-heaviness; or 10 metre-tons, solar surface-heaviness, because a ton of matter is twenty-seven times as heavy at the sun's surface as at the earth's. To do 78,000 horse-power, or 780,000 metre-tons solar surface-heaviness per hour, our weight must therefore descend at the rate of one metre in 313 hours, or about 28 metres per year.

To advance another step, still through impracticable mechanism, towards the practical method by which the sun's heat is produced, let the thread of the screw be of uniformly decreasing steepness from the surface downwards, so that the velocity of the weight, as it is allowed to descend by the turning of the screw, shall be in simple proportion to distance from the sun's centre. This will involve a uniform condensation of the material of the weight; but a condensation so exceedingly small in the course even of tens of thousands of years, that, whatever be the supposed material, metal or stone, of the weight, the elastic resistance against the condensation will be utterly imperceptible in comparison with the gravitational forces with which we are concerned. The work done per metre of descent of the top end of the weight will be just four-fifths of what it was when the thread of the screw was uniform. Thus, to do the 78,000 horse-power of work, the top end of the weight must descend at the rate of 35 metres per year: or 70 kilometres per 2,000 years.

Now let the whole surface of our cool solid sun be divided into squares, for example as nearly as may be of one square metre area each, and let the whole mass of the sun be divided into long inverted pyramids or pointed rods, each 697,000 kilometres long, with their points meeting at the centre. Let each be mounted on a screw, as already described for the long tapering weight which we first considered; and let the paddle at the top end of each screw-shaft revolve in a fluid, not now confined to a vat, but covering the whole surface of the sun to a depth of a few metres or kilometres. Arrange the viscosity of the fluid and the size of each paddle so as to let the paddle turn just so fast as to allow the top end of each pointed rod to descend at the rate of 35 metres per year. The whole fluid will, by the work which the paddles do in it, be made incandescent, and it will give out heat and light to just about the same amount as is actually done by the sun. If the fluid be a few thousand metres deep over the paddles, it would be impossible, by any of the appliances of solar physics, to see the difference between our model mechanical sun and the true sun.

To do away with the last vestige of impracticable mechanism, in which the heavinesses of all parts of each long rod are supported on the thread of an ideal screw cut on a vertical shaft of ideal matter, absolutely hard and absolutely frictionless: first, go back a step to our supposition of just one such rod and screw working in a single pit excavated down to the centre of the sun, and let us suppose all the rest of the sun's mass to be rigid and absolutely impervious to heat. Warm up the matter of the pyramidal rod to such a temperature that its material melts and experiences as much of Sir Humphry Davy's "repulsive motion" as suffices to keep it balanced as a fluid, without either sinking or rising from the position in which it was held by the thread of the screw. When the matter is thus held up without the screw, take away the screw or let it melt in its place.

We should thus have a pit from the sun's surface to his centre, of a square metre area at the surface, full of incandescent fluid, which we may suppose to be of the actual ingredients of the solar substance. This fluid, having at the first instant the temperature with which the paddle left it, would for that instant continue radiating heat just as it did when the paddle was kept moving ; but it would quickly become much cooler at its surface, and to a distance of a few metres down. Currents of less hot fluid tumbling down, and hotter fluid coming up from below, in irregular whirls, would carry the cooled fluid down from the surface, and bring up hotter fluid from below, but this mixing could not go on through a depth of very many metres to a sufficient degree to keep up anything approaching to the high temperature maintained by the paddle; and after a few hours or days, solidification would commence at the surface. If the solidified matter floats on the fluid, at the same temperature, below it, the crust would simply thicken as ice on a lake thickens in frosty weather; but if, as is more probable, solid matter, of such ingredients as the sun is composed of, sinks in the liquid when both are at the melting temperature of the substance, thin films of the upper crust would fall in, and continue falling in, until, for several metres downwards, the whole mass of mixed solid and fluid becomes stiff enough (like the stiffness of paste or of mortar) to prevent the frozen film from falling down from the surface. The surface film would then quickly thicken, and in the course of a few hours or days become less than red-hot on its upper surface, the whole pit full of fluid would go on cooling with extreme slowness until, after possibly about a million million million years or so, it would be all at the same temperature as the space to which its upper end radiates.

Let precisely what we have been considering

be done for every one of our pyramidal rods, with, however, in the first place, thin partitions of matter impervious to heat separating every pit from its four surrounding neighbours. Precisely the same series of events as we have been considering will take place in every one of the pits.

Suppose the whole complex mass to be rotating at the rate of once round in twenty-five days, which is, about as exactly as we know it, the time of the sun's rotation about its axis.

Now at the instant when the paddle stops let all the partitions be annulled, so that there shall be perfect freedom for currents to flow unresisted in any direction, except so far as resisted by the viscosity of the fluid, and leave the piece of matter, which we may now call the Sun, to himself. He will immediately begin showing all the phenomena known in solar physics. Of course the observer might have to wait a few years for sunspots, and a few quarter-centuries to discover periods of sunspots, but they would, I think I may say probably, all be there just as they are, because I think we may feel that it is most probable that all these actions are due to the sun's own substance, and not to external influences of any kind. It is, however, quite possible, and indeed many who know most of the subject think it probable, that some of the chief phenomena due to sunspots arise from influxes of meteoric matter circling round the sun.

The energy of chemical combination is as nothing compared with the gravitational energy of shrinkage, to which the sun's activity is almost wholly due. A body falling forty-six kilometres to the sun's surface *or through the sun's atmosphere*, has as much work done on it by gravity, as corresponds to a high estimate of chemical energy in the burning of combustible materials. But chemical combinations and dissociations may, as urged by Lockyer, in his book on the *Chemistry of the Sun*, just now published, be thoroughly potent determining influences on some of the features of non-uniformity of the brightness in the grand phenomena of sunspots, hydrogen flames, and corona, which make the province of solar physics. But these are questions belonging to a very splendid branch of solar science to which only allusion can be made at the present time.

What concerns us as to the explanation of sun-light and sun-heat may be summarised in two propositions :—

(1) Gigantic currents throughout the sun's liquid mass are continually maintained by fluid, slightly cooled by radiation, falling down from the surface, and hotter fluid rushing up to take its place.

(2) The work done in any time by the mutual gravitation of all the parts of the fluid, as it shrinks in virtue of the lowering of its temperature, is but little less than (so little less than that we may regard it as practically equal to) the dynamical equivalent of the heat that is radiated from the sun in the same time.

The rate of shrinkage corresponding to the present rate of solar radiation has been proved to us, by the consideration of our dynamical model, to be 35 metres on the radius per year, or one ten-thousandth of its own length on the radius per two thousand years. Hence, if the solar radiation has been about the same as at present for two hundred thousand years, his radius must have been greater by one per cent. two hundred thousand years ago than at present. If we wish to carry our calculations much farther back or forward than two hundred thousand years, we must reckon by differences of the reciprocal of the sun's radius, and not by differences simply of the radius, to take into account the change of density (which, for example, would be three per cent. for one per cent. change of the radius). Thus the rule, easily worked out according to the principles illustrated by our mechanical model, is this :—

Equal differences of the reciprocal of the radius correspond to equal quantities of heat radiated away from million of years to million of years.

Take two examples—

(1) If in past time there has been as much as fifteen million times the heat radiated from the sun as is at present radiated out in one

year, the solar radius must have been four times as great as at present.

(2) If the sun's effective thermal capacity can be maintained by shrinkage till twenty million times the present year's amount of heat is radiated away, the sun's radius must be half what it is now. But it is to be remarked that the density which this would imply, being 11·2 times the density of water, or just about the density of lead, is probably too great to allow the free shrinkage as of a cooling gas to be still continued without obstruction through overcrowding of the molecules. It seems, therefore, most probable that we cannot for the future reckon on more of solar radiation than, if so much as, twenty million times the amount at present radiated out in a year. It is also to be remarked that the greatly diminished radiating surface, at a much lower temperature, would give out annually much less heat than the sun in his present condition gives. The same considerations led Newcomb to the conclusion "That it is

hardly likely that the sun can continue to give sufficient heat to support life on the earth (such life as we now are accquainted with at least) for ten million years from the present time."

In all our calculations hitherto we have for simplicity taken the density as uniform throughout, and equal to the true mean density of the sun, being about 1·4 times the density of water, or about a quarter of the earth's mean density. In reality the density in the upper parts of the sun's mass must be something less than this, and something considerably more than this in the central parts, because of the pressure in the interior increasing to something enormously great at the centre. If we knew the distribution of interior density we could easily modify our calculations accordingly; but it does not seem probable that the correction could, with any probable assumption as to the greatness of the density throughout a considerable proportion of the sun's interior, add more than a few million years to the past of solar heat, and

what could be added to the past must be taken
from the future.

In our calculations we have taken Pouillet's
number for the total activity of solar radiation,
which practically agrees with Herschel's. Forbes[1]
showed the necessity for correcting the mode
of allowing for atmospheric absorption used by
his two predecessors in estimating the total
amount of solar radiation, and he was thus led
to a number 1·6 times theirs. Forty years later
Langley,[2] in an excellently worked out consideration of the whole question of absorption by
our atmosphere, of radiant heat of all wave-
lengths, accepts and confirms Forbes's reasoning,
and by fresh observations in very favourable
circumstances on Mount Whitney, 15,000 feet
above the sea-level, finds a number a little
greater still than Forbes (1·7, instead of Forbes's
1·6, times Pouillet's number). Thus Langley's
measurement of solar radiation corresponds to
133,000 horse-power per square metre, instead

---

[1] *Edin. New Phil. Journal,* vol. xxxvi. 1844.
[2] *American Journal of Science,* vol. xxvi. March, 1883.

of the 78,000 horse-power which we have taken, and diminishes each of our times in the ratio of 1 to 17. Thus, instead of Helmholtz's twenty million years, which was founded on Pouillet's estimate, we have only twelve millions, and similarly with all our other time reckonings based on Pouillet's results. In the circumstances, and taking fully into account all possibilities of greater density in the sun's interior, and of greater or less activity of radiation in past ages, it would, I think, be exceedingly rash to assume as probable anything more than twenty million years of the sun's light in the past history of the earth, or to reckon on more than five or six million years of sunlight for time to come.

We have seen that the sun draws on no external source for the heat he radiates out from year to year, and that the whole energy of this heat is due to the mutual attraction between his parts acting in conformity with the Newtonian law of gravitation. We have seen how an ideal mechanism, easily imagined and understood, though infinitely far from possibility of realisation, could direct the work done by mutual gravitation between all the parts of the shrinking mass, to actually generate its heat-equivalent in an ocean of white-hot liquid covering the sun's surface, and so keep it white-hot while constantly radiating out heat at the actual rate of the sun's heat-giving activity. Let us now consider a little more in detail the real forces and movements actually concerned in the process of cooling by radiation from the outermost region of the sun, the falling inwards of the fluid thus cooled, the consequent mixing up of the whole mass of the sun, the resulting diminished elastic resistance to pressure in equi dense parts, and the consequent shrinkage of the whole mass under the influence of mutual gravitation. I must first explain that this "elastic resistance to pressure" is due to heat, and is, in fact, what I have, in the present lecture, called "Sir Humphry Davy's repulsive motion" (p. 381). I called it so because Davy first used the expression "repulsive motion" to describe the fine intermolecular motions to which he and other founders of the Kinetic Theory of Heat attributed the elastic resistance to compression presented by gases and fluids.

Imagine, instead of the atoms and molecules of the various substances which constitute the sun's mass, a vast number of elastic globes, like schoolboys' marbles or billiard balls. Consider first, anywhere on our earth a few million such balls put into a room, large enough to hold a thousand times their number, with perfectly hard walls and ceiling, but with a real wooden floor; or, what would be still more convenient for our purpose, a floor of thin elastic sheet steel, supported by joists close enough together to prevent it from drooping inconveniently in any part. Suppose in the beginning the marbles to be lying motionless on the floor. In this condition they represent the atoms of a gas, as for instance, oxygen, nitrogen, or hydrogen, absolutely deprived of heat, and therefore lying frozen, or as molecular dust strewn on the floor of the containing vessel.

If now a lamp be applied below the oxygen, nitrogen, or hydrogen, the substance becoming warmed by heat conducted through the floor will rise from its condition of absolutely cold solid, or of incoherent molecular dust, and will spread as a gas through the whole inclosed space. If more and more heat be applied by the lamp the pressure of the gas outwards in all directions against the inside of the enclosing vessel will become greater and greater.

As a rude mechanical analogue to this warming of a gas by heat conducted through the floor of its containing vessel from a lamp held below it, return to our room with floor strewn with marbles, and employ workmen to go below the floor and strike its underside in a great many places vehemently with mallets. The marbles in immediate contact with the floor will begin to jump from it and fall sharply back again (like water in a pot on a fire simmering before it boils). If the workmen work energetically enough there will be more and more of commotion in the heap, till every one of the balls gets into a state of irregular vibration, up and down, or obliquely, or horizontally, but in no fixed direction; and by mutually jostling the heap swells up till the ceiling of the room prevents it from swelling any further. Suppose now the floor to become, like the walls and ceiling, absolutely rigid. The workmen may cease their work of hammering, which would now be no more availing to augment the motions of the marbles within, than would be a lamp applied outside to warm the contents of a vessel, if the vessel were made of ideal matter impermeable to heat. The marbles being perfectly elastic will continue for ever [1] flying about in their room, striking the walls

To justify this statement I must warn the reader that the ideal perfectly elastic balls which we are imagining, must be supposed somehow to have such a structure that each takes only a definite average proportion of its share of the kinetic energy of the whole multitude, so that on the average there is a constant proportion of energy in the translatory motions of the balls; the other part being the vibratory or rotational motions of the parts of each ball. For simplicity also we suppose the balls to be perfectly smooth and frictionless, so that we shall not be troubled by need to consider them as having any rotatory motions, such as real balls with real frictional collisions would acquire. The ratio of the two kinds of energy for ordinary gases, according to Clausius, to whom is due this essential contribution to the kinetic theory, is—of the whole energy, three-fifths translational to two fifths vibrational.

and floor and ceiling and one another, and remaining in a constant average condition of denser crowd just over the floor and less and less dense up to the ceiling.

In this constant average condition the average velocity of the marbles will be the same all through the crowd, from ceiling to floor, and will be the same in all directions, horizontal, or vertical, or inclined. The continually repeated blows upon any part of the walls or ceiling will in the aggregate be equivalent to a continuous pressure which will be in simple proportion to the average density of the crowd at the place. The diminution of pressure and density from the floor upwards will be precisely the same as that of the density and pressure of our atmosphere, calculated on the supposition of equal temperature at all heights, according to the well-known formula and tables for finding heights by the barometer.

In reality the temperature of the atmosphere is not uniform from the ground upwards, but diminishes at the rate of about 1° C for every 162 metres of vertical ascent in free air, undisturbed by mountains, according to observations made in balloons by the late Mr. Welsh, of Kew, through a large range of heights. This diminu tion of temperature upwards in our terrestrial atmosphere is most important and suggestive in respect to the constitution of the solar atmosphere, and not merely of the atmosphere or outer shell of the sun, but of the whole interior fluid mass with which it is continuous. The two cases have so much in common, that there is in each case loss of heat from the outer parts of the atmosphere by radiation into space, and that in consequence circulating currents are produced through the continuous fluid, by which a thorough mixing up and down is constantly performed. In the case of the terrestrial atmosphere the lowest parts receive, by contact, heat from the solid earth, warmed daily by the sun's radiation. On the average of night and day, as the air does not become warmer on the whole, it must radiate out into space as much heat as all that it gets, both from the earth by contact, and by radiation of heat from the earth, and by intercepted radiation

from the sun on its way to the earth. In the case
of the sun the heat radiated from the outer parts
of the atmosphere is wholly derived from the
interior. In both cases the whole fluid mass is
kept thoroughly mixed by currents of cooled fluid
coming down, and of warmer fluid rising to take
its place and to be cooled and descend in its turn.

Now it is a well-known property of gases and
of fluids generally (except some special cases, as
that of water within a few degrees of its freezing
temperature, in which the fluid under constant
pressure contracts with rise of temperature) that
condensations and rarefactions, effected by aug-
mentations and diminutions of pressure from
without, produce elevations and lowerings of
temperature in circumstances in which the gas is
prevented from either taking heat from or giving
heat to any material external to it. Thus a
quantity of air or other gas taken at ordinary
temperature (say 15° C. or 59° F.) and expanded
to double its bulk becomes 71° C. cooler; and if
the expansion is continued to thirty-two times
its original bulk it becomes cooled 148° further,

or down to about 200° C. below the temperature of freezing water, or to within 73° of absolute cold. Such changes as these actually take place in masses of air rising in the atmosphere to heights of eight or nine kilometers, or of twenty or twenty-five kilometers. Corresponding differences of temperature there certainly are throughout the fluid mass of the sun, but of very different magnitudes because of the twenty-sevenfold greater gravity at the sun's surface, the vastness of the space through which there is free circulation of fluid, and last, though not least, the enormously higher temperature of the solar fluid than of the terrestrial atmosphere at points of equal density in the two. This view of the solar constitution has been treated mathematically with great power by Mr. J. Homer Lane, of Washington, U.S.A., in a very important paper read before the National Academy of Sciences of the United States in April, 1869, and published with further developments in the *American Journal of Science*, for July, 1870. Mr. Lane, by strict mathematical treatment finds the law of distribution of density and temperature all through a globe of homogeneous gas left to itself in space, and losing heat by radiation outwards so slowly that the heat-carrying currents produce but little disturbance from the globular form.

One very remarkable and important result which he finds is, that the density at the centre is about twenty [1] times the mean density ; and this, whether the mass be large or small, and whether of oxygen, nitrogen, or hydrogen, or other substance ; pro vided only it be of one kind of gas throughout, and that the density in the central parts is not too great to allow the condensation to take place, according to the ordinary gaseous law of density, in simple proportion to pressure for the same temperatures. We know this law to hold with somewhat close accuracy for common air, and for each of its two chief constituents, oxygen and nitrogen, separately, and for hydrogen, to densities of about two hundred times their densities at our ordinary atmospheric pressure. But when the compressing force is sufficiently increased, they

[1] Working out Lane's problem independently, I find 22½ as very nearly the exact number.

all show greater resistance to condensation than according to the law of simple proportion, and it seems most probable that there is for every gas a limit beyond which the density cannot be increased by any pressure however great. Lane remarks that the density at the centre of the sun would be "nearly one-third greater than that of the metal platinum," if the gaseous law held up to so great a degree of condensation for the ingredients of the sun's mass; but he does not suggest this supposition as probable, and he no doubt agrees with the general opinion that in all probability the ingredients of the sun's mass, at the actual temperatures corresponding to their positions in his interior, obey the simple gaseous law through but a comparatively small space inwards from the surface, and that in the central regions they are much less condensed than according to that law. According to the simple gaseous law, the sun's central density would be thirty-one times that of water; we may assume that it is in all probability much less than this, though considerably greater than the mean density, 1·4.

This is a wide range of uncertainty, but it would be unwise at present to narrow it, ignorant as we are of the main ingredients of the sun's whole mass, and of the laws of pressure, density, and temperature, even for known kinds of matter, at very great pressures and very high temperatures.

The question, Is the sun becoming colder or hotter? is an exceedingly complicated one, and, in fact, either to put it or to answer it is a paradox, unless we define exactly where the temperature is to be reckoned. If we ask, How does the temperature of equi-dense portions of the sun vary from age to age? the answer certainly is that the matter of the sun of which the density has any stated value, for example, the ordinary density of our atmosphere, becomes always less and less hot, whatever be its place in the fluid, and whatever be the law of compression of the fluid, whether the simple gaseous law or anything from that to absolute incompressibility. But the distance inwards from the surface at which a constant density is to be found diminishes with shrinkage, and thus it may be that at constant

D D

depths inwards from the bounding surface the temperature is becoming higher and higher. This would certainly be the case if the gaseous law of condensation held throughout, but even then the effective radiational temperature, in virtue of which the sun sheds his heat outwards, might be becoming lower, because the temperatures of equidense portions are clearly becoming lower under all circumstances.

Leaving now these complicated and difficult questions to the scientific investigators who are devoting themselves to advancing the science of solar physics, consider the easily understood question, What is the temperature of the centre of the sun at any time, and does it rise or fall as time advances? If we go back a few million years, to a time when we may believe the sun to have been wholly gaseous to the centre, then certainly the central temperature must have been augmenting; again, if, as is possible though not probable at the present time, but may probably be the case at some future time, there be a solid nucleus, then certainly the central temperature

would be augmenting, because the conduction of
heat outwards through the solid would be too
slow to compensate the augmentation of pressure
due to augmentation of gravity in the shrinking fluid
around the solid. But at a certain time in the
history of a wholly fluid globe, primitively rare
enough throughout to be gaseous, shrinking under
the influence of its own gravitation and its radiation of heat outwards into cold surrounding space,
when the central parts have become so much
condensed as to resist further condensation greatly
more than according to the gaseous law of simple
proportions, it seems to me certain that the early
process of becoming warmer, which has been
demonstrated by Lane, and Newcombe, and Ball,
must cease, and that the central temperature
must begin to diminish on account of the cooling by radiation from the surface, and the mixing
of the cooled fluid throughout the interior.

Now we come to the most interesting part of
our subject—the early history of the Sun. Five
or ten million years ago he may have been about
double his present diameter and an eighth of his

present mean density, or 175 of the density of water; but we cannot, with any probability of argument or speculation, go on continuously much beyond that. We cannot, however, help asking the question, What was the condition of the sun's matter before it came together and became hot? It may have been two cool solid masses, which collided with the velocity due to their mutual gravitation; or, but with enormously less of probability, it may have been two masses colliding with velocities considerably greater than the velocities due to mutual gravitation. This last supposition implies that, calling the two bodies A and B for brevity, the motion of the centre of inertia of B relatively to A, must, when the distances between them was great, have been directed with great exactness to pass through the centre of inertia of A; such great exactness that the rotational momentum, or "moment of momentum,"[1] after collision was no more than

[1] This is a technical expression in dynamics which means the importance of motion relatively to revolution or rotation round an axis. Momentum is an expression given about a hundred and fifty years ago (when mathematicians and other learned men spoke and

to let the sun have his present slow rotation
when shrunk to his present dimensions. This
exceedingly exact aiming of the one body at
the other, so to speak, is, on the dry theory of
probability, exceedingly improbable. On the
other hand, there is certainty that the two bodies
A and B, at rest in space, if left to themselves
undisturbed by other bodies and only influenced
by their mutual gravitation, shall collide with
direct impact, and therefore with no motion of
their centre of inertia, and no rotational momentum
of the compound body after the collision. Thus
we see that the dry probability of collision between
two neighbours of a vast number of mutually
attracting bodies widely scattered through space
is much greater if the bodies be all given at rest,

wrote Latin) to signify translational importance of motion. Moment
of a couple, moment of a magnet, moment of inertia, moment of
force round an axis, moment of momentum round an axis, and
corresponding verbal combinations in French and German, are
expressions which have been introduced within the last sixty years
(by scientists speaking, as now, each his own vernacular) to signify
the importance of the special subject referred to in each case. The
expression "moment of momentum" is highly valuable and convenient in dynamical science, and it constitutes a curious philological
monument of scientific history.

than if they be given moving in any random
directions and with any velocities considerable in
comparison with the velocities which they would
acquire in falling from rest into collision.   In
this connection it is most interesting to know from
stellar astronomy, aided so splendidly as it has
recently been by the spectroscope, that the relative
motions of the visible stars and our sun are
generally very small in comparison with the
velocity (612 kilometres per second) which a body
would acquire in falling into the sun, and are comparable with the moderate little velocity (29·5 kilometres per second) of the earth in her orbit round
the sun.

To fix the ideas, think of two cool solid globes,
each of the same mean density as the earth and
of half the sun's diameter, given at rest, or nearly
at rest, at a distance asunder equal to twice the
earth's distance from the sun.   They will fall together and collide in exactly half a year.   The
collision will last for about half an hour, in the
course of which they will be transformed into a
violently agitated incandescent fluid mass flying

outward from the line of the motion before the collision, and swelling to a bulk several times greater than the sum of the original bulks of the two globes.[1] How far the fluid mass will fly out all round from the line of collision it is impossible to say. The motion is too complicated to be fully investigated by any known mathematical method; but with sufficient patience a mathematician might be able to calculate it with some fair approximation to the truth. The distance reached by the extreme circular fringe of the fluid mass would probably be much less than the distance fallen by each globe before the collision, because the translational motion of the molecules constituting the heat into which the whole energy

[1] Such incidents seem to happen occasionally in the universe. Laplace says "Some stars have suddenly appeared, and then disappeared, after having shone for several months with the most brilliant splendour. Such was the star observed by Tycho Brahe in the year 1572, in the constellation Cassiopeia. In a short time it surpassed the most brilliant stars, and even Jupiter itself. Its light then waned away, and finally disappeared sixteen months after its discovery. Its colour underwent several changes; it was at first of a brilliant white, then of a reddish yellow, and finally of a lead-coloured white, like to Saturn." (Harte's translation of Laplace's *System of the World.* Dublin, 1830.)

of the original fall of the globes becomes transformed in the first collision, takes probably about
three-fifths of the whole amount of that energy.
The time of flying out would probably be less
than half a year, when the fluid mass must begin
to fall in again towards the axis. In something
less than a year after the first collision the fluid
will again be in a state of maximum crowding
round the centre, and this time probably even
more violently agitated than it was immediately
after the first collision ; and it will again fly outward, but this time axially towards the places
whence the two globes fell. It will again fall
inwards, and after a rapidly subsiding series of
quicker and quicker oscillations it will subside,
probably in the course of two or three years, into
a globular star of about the same mass, heat,
and brightness, as our present sun, but differing from him in this, that it will have no
rotation.

We supposed the two globes to have been at
rest when they were let fall from a mutual distance
equal to the diameter of the earth's orbit. Sup-

pose, now, that instead of having been at rest they had been moving transversely in opposite directions with a relative velocity of two (more exactly 1·89) metres per second. The moment of momentum of these motions round an axis through the centre of gravity of the two globes perpendicular to their lines of motion, is just equal to the moment of momentum of the sun's rotation round his axis. It is an elementary and easily proved law of dynamics that no mutual action between parts of a group of bodies, or of a single body, rigid, flexible, or fluid, can alter the moment of momentum of the whole. The transverse velocity in the case we are now supposing is so small that none of the main features of the collision and of the wild oscillations following it, which we have been considering, or of the magnitude, heat, and brightness of the resulting star, will be sensibly altered ; but now, instead of being rotationless, it will be revolving once round in twenty-five days and so will be in all respects like to our sun.

If instead of being at rest initially, or moving with the small transverse velocities we have been considering, each globe had a transverse velocity of three-quarters (or anything more than 71) of a kilometre per second, they would just escape collision, and would revolve in ellipses round their common centre of inertia in a period of one year, just grazing each other's surface every time they came to the nearest points of their orbits.

If the initial transverse velocity of each globe be less than, but not much less than, ·71 of a kilometre per second, there will be a violent grazing collision, and two bright suns, solid globes bathed in flaming fluid, will come into existence in the course of a few hours, and will commence revolving round their common centre of inertia in long elliptic orbits in a period of a little less than a year. Tidal interaction between them will diminish the eccentricities of their orbits, and if continued long enough will cause the two to revolve in circular orbits round their centre of inertia with a distance between their surfaces equal to 6·44 diameters of each.

Suppose now, still choosing a particular case to fix the ideas, that twenty-nine million cold, solid globes, each of about the same mass as the moon, and amounting in all to a total mass equal to the sun's, are scattered as uniformly as possible on a spherical surface of radius equal to one hundred times the radius of the earth's orbit, and that they are left absolutely at rest in that position. They will all commence falling towards the centre of the sphere, and will meet there in two hundred and fifty years, and every one of the twenty-nine million globes will then, in the course of half an hour, be melted, and raised to a temperature of a few hundred thousand or a million degrees centigrade. The fluid mass thus formed will, by this prodigious heat, be exploded outwards in vapour or gas all round. Its boundary will reach to a distance considerably less than one hundred times the radius of the earth's orbit on first flying out to its extreme limit. A diminishing series of out-and-in oscillations will follow, and the incandescent globe thus contracting and expanding alternately, in the course it may be of three or four hundred years, will settle to a radius of forty[1] times the

---

[1] The radius of a steady globular gaseous nebula of any homo-geneous gas is 40 per cent. of the radius of the spheric surface

radius of the earth's orbit. The average density of the gaseous nebula thus formed would be $(215 \times 40)^{-3}$, or one six hundred and thirty-six thousand millionth of the sun's mean density; or one four hundred and fifty-four thousand millionth of the density of water; or one five hundred and seventy millionth of that of common air at an ordinary temperature of $10°$ C. The density in its central regions, sensibly uniform through several million kilometres, is (see note on p. 399) one twenty thousand millionth of that of water; or one twenty-five millionth of that of air. This exceedingly small density is nearly six times the density of the oxygen and nitrogen left in some of the receivers exhausted by Bottomley in his experimental measurements of the amount of heat emitted by pure radiation from highly heated bodies. If the substance were oxygen, or nitrogen, or other gas or mixture of gases simple or compound, of specific density equal to the specific density of our air, the central temperature would

from which its ingredients must fall to their actual positions in the nebula to have the same kinetic energy as the nebula has.

be 51,200° C., and the average translational velocity of the molecules 6·7 kilometres per second, being $\sqrt{\frac{3}{7}}$ of 10·2, the velocity acquired by a heavy body falling unresisted from the outer boundary (of 40 times the radius of the earth's orbit) to the centre of the nebulous mass.

The gaseous nebula thus constituted would in the course of a few million years, by constantly radiating out heat, shrink to the size of our present sun, when it would have exactly the same heating and lighting efficiency, but no notion of rotation.

The moment of momentum of the whole solar system is about eighteen times that of the sun's rotation ; seventeen-eighteenths being Jupiter's and one-eighteenth the sun's, the other bodies being not worth taking into account in the reckoning of moment of momentum.

Now instead of being absolutely at rest in the beginning, let the twenty-nine million moons be given each with some small motion, making up in all an amount of moment of momentum about a certain axis, equal to the moment of momentum of the solar system which we have just been con-

sidering; or considerably greater than this, to allow
for effect of resisting medium. They will fall to
gether for two hundred and fifty years, and though
not meeting precisely in the centre as in the first
supposed case of no primitive motion, they will,
two hundred and fifty years from the beginning,
be so crowded together that there will be myriads
of collisions, and almost every one of the twenty-
nine million globes will be melted and driven into
vapour by the heat of these collisions. The vapour
or gas thus generated will fly outwards, and after
several hundreds or thousands of years of outward
and inward oscillatory motion, may settle into an
oblate rotating nebula extending its equatorial
radius far beyond the orbit of Neptune, and with
moment of momentum equal to or exceeding the
moment of momentum of the solar system. This
is just the beginning postulated by Laplace for his
nebular theory of the evolution of the solar system ;
which, founded on the natural history of the stellar
universe, as observed by the elder Herschel, and
completed in details by the profound dynamical
judgment and imaginative genius of Laplace, seems

converted by thermodynamics into a necessary truth, if we make no other uncertain assumption than that the materials at present constituting the dead matter of the solar system have existed under the laws of dead matter for a hundred million years. Thus there may in reality be nothing more of mystery or of difficulty in the automatic pro gress of the solar system from cold matter diffused through space, to its present manifest order and beauty, lighted and warmed by its brilliant sun, than there is in the winding up of a clock[1] and letting it go till it stops. I need scarcely say that the beginning and the maintenance of life on the earth is absolutely and infinitely beyond the range of all sound speculation in dynamical science. The only contribution of dynamics to theoretical biology is absolute negation of automatic commencement or automatic maintenance of life.

I shall only say in conclusion :—Assuming the sun's mass to be composed of materials which

---

[1] Even in this, and all the properties of matter which it involves, there is enough, and more than enough, of mystery to our limited understanding. A watch-spring is much farther beyond our understanding than is a gaseous nebula.

were far asunder before it was hot, the immediate antecedent to its incandescence must have been either two bodies with details differing only in proportions and densities from the cases we have been now considering as examples ; or it must have been some number more than two—some finite number—at the most the number of atoms in the sun's present mass, a finite number (which may probably enough be something between 4×10^{57} and 140×10^{57}) as easily understood and imagined as numbers 4 or 140. The immediate antecedent to incandescence may have been the whole constituents in the extreme condition of subdivision— that is to say, in the condition of separate atoms ; or it may have been any smaller number of groups of atoms making minute crystals or groups of crystals—snowflakes of matter, as it were ; or it may have been lumps of matter like a macadamising stone ; or like this stone [1] (Fig. 50), which you

[1] These three meteorites are in the possession of the Hunterian Museum of the University of Glasgow, and the wood-cuts, Figs. 50, 51, and 52, have been executed from the actual specimens kindly lent for the purpose by the keeper of the museum, Professor Young The specimen represented by Fig. 50 is contained in the Hunterian

might mistake for a macadamising stone, but which was actually travelling through space till it

FIG. 50.
←··· .··············5 centimetres··········· ···→

fell on the earth at Possil, in the neighbourhood of Glasgow, on April 5th, 1804 ; or like that (Fig. 51) which was found in the Desert of Atacama, in

collection, that by Fig. 51 in the Eck collection, and that by Fig. 52 in the Lanfine collection—the scale of dimensions is shown for each. It may be remarked that Fig. 51 represents a section of the meteorite taken in the plane of the longest of three rectangular axes; the bright markings being large and well-formed crystals of olivine, embedded in a matrix of iron. In Fig. 52 is depicted the beautiful Widmanstätten marking characteristic of all meteoric iron, and so well shown in the well-known Lenarto meteorite.

South America, and is believed to have fallen there from the sky—a fragment made up of iron and stone, which looks as if it has solidified from

FIG. 51.

←···13½ centimetres···→

a mixture of gravel and melted iron in a place where there was very little of heaviness; or this splendidly crystallised piece of iron (Fig. 52), a slab cut out of the celebrated aërolite which fell at

Lenarto, in Hungary ;[1] or this wonderfully-shaped specimen (of which two views are given in Figs. 53 and 54), a model of the Middlesburgh meteorite (kindly given me by Professor A. S. Herschel), having corrugations showing how its

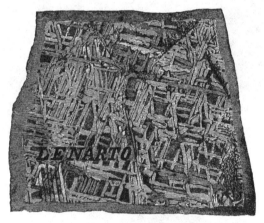

Fig. 52.

←.............................9¼ centimetres.............................→

melted matter has been scoured off from the front part of its surface, in its final rush through the earth's atmosphere when it was seen to fall on March 14, 1881, at 3.35 P.M.

[1] See footnote, pp. 416, 417.

E E 2

FIG. 53.

←..................................15 centimetres..................................→

FIG. 54.

For the theory of the sun it is indifferent which
of these varieties of configurations of matter may
have been the immediate antecedent of his
incandescence, but I can never think of these
material antecedents without remembering a
question put to me thirty years ago by the late
Bishop Ewing, Bishop of Argyll and the Isles:
" Do you imagine that piece of matter to have
been as it is from the beginning; to have been
created as it is, or to have been as it is through
all time till it fell on the earth ?" I had told
him that I believed the sun to be built up of
meteoric stones, but he would not be satisfied till
he knew or could imagine what kind of stones.
I could not but agree with him in feeling it impossible to imagine that any one of such
meteorites as those now before you has been
as it is through all time, or that the materials of
the sun were like this for all time before they came
together and became hot. Surely this stone has
an eventful history, but I shall not tax your
patience by trying just now to trace it conjec-
turally. I shall only say that we cannot but agree

with the common opinion which regards meteorites as fragments broken from larger masses, and we cannot be satisfied without trying to imagine what were the antecedents of those masses.

ELECTRICAL MEASUREMENT.

[*Address before Section of Mechanics at the Conferences held in connection with the Special Loan Collection of Scientific Apparatus at the South Kensington Museum, May 17th, 1876.*]

THE beginnings of electrical measurements, are, I believe, the measurements of Robinson in Edinburgh, and of Coulomb in Paris of electrostatic forces. The great results which followed from those measurements illustrated how important is accurate measurement in promoting thorough scientific knowledge in any branch of physical science. The earlier electricians merely describe phenomena — attractions and repulsions and flashes and sparks—and the nearest approach to measurement which they gave us, was the length of the spark under certain circumstances ;

the other circumstances on which the length of
the spark might depend being left unmeasured.
By Robinson's and Coulomb's experiments was
established the law of electrostatic force, according
to which two small bodies, each electrified with
a constant quantity of electricity, exercise a mutual
force of attraction or repulsion, according as the
electricity is similar or dissimilar, and which varies
inversely as the square of the distance, when the
distance between the two bodies is varied.

In physical science generally, measurement involves one or other of two methods :—a method
of adjustment to a zero, or what is called a *null*
method ; and a method of measuring some continuously varying quantity. This second branch
of measurement was illustrated in Coulomb's and
Robinson's experiments, where the law was determined according to which the electric force
varies, when the distance between the mutually
influencing bodies varies continuously. The other
mode of experimenting in connection with measurement is illustrated by another exceedingly important phenomenon, bearing upon electrical theory

and that is the evanescence of electrical force in the interior of a conductor. Both kinds of measurements were practised by Cavendish in a very remarkable manner, and I look forward with great expectation to the results we are soon to have of Cavendish's work. One most interesting result which will follow from the Cavendish laboratory in Cambridge — from its director Professor Clerk Maxwell—and from the relationship thus established between the physical laboratory of the University of Cambridge, and its director on the one hand, and the munificent founder of the institution, the Duke of Devonshire, on the other hand, is this : the Cavendish manuscripts which still remain in that family, being, I believe, at present in the possession of the Duke of Devonshire, have been by him put into the hands of Professor Clerk Maxwell for the purpose of having published either the whole, or such extracts from them as may be found to be of scientific interest at the present day. The whole of them, no doubt, had great scientific interest at one time. A large part of these manu scripts, I believe, will be found to be excessively interesting even now, and from something I heard a few days ago from Professor Clerk Maxwell, when he was here on the opening day of this Conference, I learnt that much more than was ever imagined is to be found in these manuscripts, and particularly that in them has been found a whole system of electrical measurements worked out, from the measurement of electrostatic capacity. The very idea of measuring electrostatic capacity in a definite scientific way is, as it now turns out, due to Cavendish. A great many years ago, in 1846 or 1847, when the Cavendish manuscripts were in the hands of Sir Wm. Snow Harris, at Plymouth, I myself found one paper, out of a box full of unsorted manuscripts, which startled me exceedingly. It contained the description of an experiment and its result, measuring the electrostatic capacity of an insulated circular disc. That is one of the cases in which the theory founded by Robinson and Coulomb as developed in the hands of mathematicians who followed, allowed the result to be calculated *a priori*, and I found the result agreed within, if I remember rightly, one-half per cent. of Cavendish's measurements. When I mention these cases of the measurement of electrical force by Coulomb and Robinson, which has led to the true law of force and of the measurement of electrostatic capacity, a subject which is the least known generally, and held to be the most difficult, I have said enough to show that we must not in this century claim all the credit of being the founders of electrical measurement.

The other chief method of experimenting in connection with measurement to which I have referred is illustrated also by Cavendish's writings, that is the adjustment to a zero. It is very curious, that while Coulomb and Robinson by direct measurement of a continuously varying quantity discovered the law of the inverse square of the distance, Cavendish, quite independently, pointed out by very subtle mathematical reasoning that the law must either be the inverse square of the distance, or must vary in a determinate manner from the law of the inverse square of the distance if in a certain case, which he defined, either a perfect zero of electric force is observed, or if instead of a perfect zero any particular amount of electric force is observed. It is quite clear from Cavendish's writings that he believed that perfect zero would be found when the experiment should be made, but with a caution characteristic of the man and also proper to his position as an accurate philosopher and mathematician he never would state the law absolutely. He had that scrupulous conscientiousness which prevented him from guessing at the conclusion at which no doubt he himself had arrived. His mind was probably a great deal quicker than are many other minds in which the conclusion is jumped at and given as if it were proved, but he conscientiously avoided stating it as a conclusion, and held it over until exact measurement should prove whether or not it was justified by experiment.

The subject of measurement in this case of the null method pointed out by Cavendish was this. If in the interior of a hollow electrified conductor the electrostatic force upon a small insulated

and electrified body is exactly zero, then the law
of variation of the electric force must be according
to the inverse square of the distance. On the
other hand, if a certain attraction of a small
positively electrified body towards the sides of
the supposed hollow electrified conductor is
observed, then the force varies according to a
law of greater variation than according to the
inverse square of the distance ; and *vice versa* if a
small body electrified in the opposite way to the
electrification of the conductor seems to be repelled
from the sides, then the law of diminution of force
with the distance will be something less than would
be calculated according to the inverse square of
the distance. The case supposed is an insulated
electrified body—an infinitely small body—charged
with electricity opposite to that of the electrified
body. If this small body, then, put into the
interior as a test, exhibits attraction towards the
sides, the law of variation of the force will
show a greater increase than according to the
inverse square of the distance, and *vice versâ*.
It was left for Faraday to make with accuracy

the concluding experiment which crowned Cavendish's theory. Faraday found by the most thoroughly searching investigation that the electrical force in the circumstances supposed was zero, and supplied the minor proposition of Cavendish's syllogism. Therefore the law of electrostatic force varies with the inverse square of the distance. This result was obtained with far less searching accuracy by Coulomb and Robinson, because their method did not admit of the same searching accuracy. On this law is founded the whole system of electrostatic measurement in absolute measure. Mathematical theory lays down the proper electrostatic unit—that quantity of electricity which, if a quantity equal to it is possessed by each of two bodies, those two bodies act and react upon one another with unit force at unit distance. On this is founded the system of absolute measurement in electrostatics.

Cavendish's other experiments, and series of experiments—because I believe Professor Clerk Maxwell is to edit a whole series of experiments measuring electrostatic quantities — led to the general system of electrostatic measurement in absolute measure.

But now there is another great branch of electrical measurement, and that is the measurement of electro-magnetic phenomena. Our elementary knowledge of electrostatics was complete, with the exception of this minor proposition of Cavendish's syllogism, and of the great physical discovery by Faraday of the peculiar inductive quality known as the electrostatic inductive capacity of dielectrics. With these two exceptions the whole theory of electrostatics was completed in the last century. It was merely left for us to work out the mathematical conclusions from the theory of Cavendish, Coulomb, and Robinson ; and it was not until after the end of the last century that the existence of electro-magnetic force became known. Orsted made the great discovery in 1820 of the mutual connection between a magnet and a wire in which an electrical current is flowing, and the remarkable developments which were very speedily given to that discovery by Ampere, led to the foundation of the other great branch of electrical science, and pointed to the subject of electro-magnetic measure ment, upon which I must now say a word or two.

I think the principles of the mathematical theory of the mutual interaction between one another, of wires carrying electric currents, and again their mutual action upon magnets, was fully laid down by Ampère in his development of Orsted's discovery The working out of the accurate measurement of currents, and generally of the system of measure ment founded on these principles, was done al together in Germany. The great work of Gauss and Weber on terrestrial magnetism belongs strictly to this subject. I believe Gauss first laid down the system of absolute measurement for magnetic force. The definitions and mathematical theory of Poisson and Coulomb as to magnetic polarity, and the theory of magnetic force founded on it, was worked out practically by Gauss, and made the foundation of the whole system of mag netic measurement now followed in our magnetic observatories. This was an immense step in science, and one of great importance, giving, not merely definite measurement, but measurement on a certain absolute basis, which, even if all the instruments by which the measurements were made were destroyed, would still enable us to get perfectly definite results. The absolute system of units was, for the first time in physical science, worked out in consequence of Gauss' foundation of the system for terrestrial magnetism. That, then, is really the beginning of absolute measurement in magnetic science, and in electro-magnetic, and electrostatic science. Gauss and Weber carried on together the work for terrestrial magnetism, and Weber carried on by himself I believe, during Gauss' lifetime and also after his death—the system of absolute measurement in electrostatics. One most interesting result, brought out by Weber, is that the electric resistance of a wire, in respect of electric currents carried by it, is to be measured in terms of certain absolute units, which lead us to a statement of velocity in units of length per unit of time, as the proper statement for the electro-magnetic measure of the resistance of a wire. It would

F F

take too long, to occupy your attention on matters of detail, and to explain minutely how it is that resistance is to be measured by a velocity. It seems curious, but you will form a very general idea of it in this way. Suppose you have two vertical copper bars and a little transverse horizontal bar placed so as to press upon those two bars. Let the plane of those two bars be perpendicular to the magnetic meridian, and place a little transverse bar, like one step of a ladder, across the two vertical bars Let this bar be moved rapidly upwards; being moved across the line of the horizontal component of the earth's magnetic force, it will, according to one of Faraday's discoveries, experience an in ductive effect, according to which one end of it will become positively electrified, and the other negatively. Now, let the two bars upon which this presses be connected together: then the tendency I have spoken of will give rise to a current. That current may be made, as in Orsted's discovery, to cause the deflection of a galvanometer needle. Now, you will see how resistance may be measured by velocity. Let the velocity of the motion of this little bar, moved upwards in the manner I have described, be such as to produce in the galvanometer a deflection of exactly 45° Then the velocity, which gives that deflection, measures the resistance in the circuit, provided always the galvanometer be arranged to fulfil a certain definite condition as to dimensions. The essential point of this statement is that the result is independent of the magnitude of the horizontal force of the earth's magnetism. The galvanometer needle is directed by the horizontal magnetic force of the earth. Let us suppose that to be doubled; the directing force on the needle is doubled, but the inductive effect is doubled also, and, therefore, the same velocity which causes the needle of the galvanometer to be deflected 45° with one amount of magnetic force of the earth, will cause the needle to be deflected by the same number of degrees, with a different amount of magnetic force of the earth. Thus, inde pendently of any absolute measurement of the terrestrial magnetic force, we get a certain velocity which gives a certain result. Thus it is that a velocity is the proper measure of the resistance of a metallic circuit to the flow of an electric current through it.

Going now to electrostatics,—the resistance of an imperfect insulator to the transmission of electricity along it, may be measured in a curious manner in connection with the velocity. It may be measured by the reciprocal of a velocity, or in other words, the conducting power of a wire may be measured, with reference to the electrostatic phenomena, by a velocity. Thus, imagine a globe in the centre of this room, at a great distance from the walls. Imagine that globe to be two metres in diameter—one metre in radius —and let it be electrified, and hung on a fine silk thread, perfectly dry, so as to insulate perfectly. There we have a perfectly insulated globe in the middle of this room. Now if you apply one end of an excessively fine wire, say a wire one ten-thousandth of an inch in diameter, to the globe, and bring the other end of the wire to a plate of metal connected with the walls of the room—or you may suppose the walls of the room to be metallic, so that we may have no confusion owing to the imperfect conductors—then by means of this very fine wire connecting the insulated globe with the walls of the room, the globe instantly loses its electricity. By *instantly*, I mean in such a short time as it would be impossible to measure by any method we could apply—I mean a time as small as, say, one-millionth of a second—the globe would lose its electricity, if we had connected it to the walls of the room by ten or twenty yards of the finest wire we could imagine. Now suppose the wire a million times finer (if we can suppose that) than we can apply, the same thing would happen, but in a correspondingly longer time. Or take a cotton thread, and suspend by means of it such a globe as I have been imagining, surrounded with metallic walls ; that moist cotton thread will gradually diselectrify it ; in a quarter of a minute the globe will have lost perhaps half its electricity in another quarter half of the remainder, and so on. If the resistance of the conductor I have supposed is constant, the loss will follow the compound interest law—so much per cent. of the charge will be lost per second. Imagine now a conductor, of perfectly constant resistance, to be put between the ideal globe and the supposed metallic walls of the room, and imagine the globe to be connected with one of these electrometers of which I shall say a word in conclusion by an excessively fine wire going into the instrument, and suppose the electrometer to indicate a certain degree of *potential,* as we now call that subject of electric measurement really discovered by Cavendish in his measurement of electric capacity. Now suppose that we are measuring the electric value—the potential—of the charge in the globe by an electrometer, then we shall see the electrometer indications decreasing—the potential gradually going down—according to the logarithmic, or compound interest law, in the circumstances I have supposed. But instead of this being carried out, let us suppose the following conditions, which we can imagine, although it would be impossible for any mechanician to execute them. Let the globe by some imaginary means be capable of becoming gradually diminished in its diameter. Suppose, in the first place, the insulation to be exceedingly perfect, and the resistance of the conducting wire to be therefore enormously great, so that in the course of a minute or two there is but little loss of potential. Now let this globe, which is supposed to be shrinkable or extendable at pleasure, be shrunk from the metre radius to 90 centimetres radius, what will the effect be? The effect will be that the potential will increase in the ratio of 90 to 100. Shrink the globe to half its dimensions the potential will be double, and so on. That follows from the result of the mathematical theory that the electrostatic capacity of a globe is numerically equal to its radius. Now, while the globe is charged, let its radius be diminished and let it shrink at such a speed that the potential shall remain constant. There, then, you can imagine a globe losing a constant quantity of electricity per unit of time, because it is kept now at a constant potential, and a globe kept by this wonderful shrinking mechanism at a constant potential will lose a constant quantity of electricity per unit of time, losing in equal times equal quantities. In that globe going on shrinking and shrinking so as to keep a constant potential, the velocity with which the surface approaches the centre measures the conducting power of the wire in absolute electrostatic measurement. So, then, we have the very curious result that according to the electrostatic law of the phenomena we can measure, in terms of electrostatic principles, the conducting power of a wire by a velocity. Although I have put an altogether ideal case to you, it would be very wrong for me to allow you to suppose that this is an ideal kind of measurement; in point of fact, we measure regularly in electrostatic measurement the capacity of the Leyden jars in that way, and in future when any one goes to buy a Leyden jar of an optician, let him tell the optician to give him a jar of one or two metres capacity or whatever it may be, and require him to find out how to produce it. I give that as a hint to any one interested in electrostatic apparatus, or in

the furnishing of laboratories. There is no likeli-
hood that the optician will understand what is
meant, but perhaps if you teach him a little he
will soon come to understand it, and I hope in ten
years hence, in every optician's shop where Leyden
jars are sold, there will be a label put on each
jar telling that the capacity is so many centimetres. It could be done to-morrow. We have
all the means of doing it, only all have not the
knowledge.

The relation between electrostatic measurement
and electro-magnetic measurement is very interesting, and here from the supposed uninterest-
ing realms of minute and accurate measurement
we are led to the depths of science, and to
look at the great things of Nature. These old
measurements of Weber led to an approximate determination of the particular velocity, " v ", at which
the electro-magnetic resistance is numerically
equal to the electrostatic conducting power of
a wire. The particular degree of resistance of a
wire which shall be such that the velocity which
measures the resistance in electro-magnetic

measure shall be the same as the velocity which measures the conducting power in electrostatic measure, was worked out by Weber, and he found that velocity, "v", to be just about 300,000 kilometres per second. I unhappily have British statute miles in my mind, through the misfortune of having been born thirty years too soon, and I remember the velocity of light in British statute miles. That used to be considered about 192,000 miles per second, but more recent observations have brought it down to about 187,000. Now I think the equivalent of that in metres is about 300,000 kilometres per second, and that was the number found by Weber for "v". Professor Clerk Maxwell gave a theory leading towards a dynamical theory of magnetism, part of which suggested to him that the velocity for which the one measure is equal to the other in the manner I have explained should be the velocity of light. This brilliant suggestion has attracted great attention, and has become an object of intense interest, not merely for the sake of accurate electro-magnetic and electrostatic measuring—the measuring with great accuracy the relation between electrostatic and electro-magnetic units—but also in connection with physical theory. It seems, up to the present time, that the more accurate such an experiment becomes, the more nearly does the result approach to being equal to the velocity of light, but still we must hold opinion in reserve before we can say that. The result has to be much closer than has been shown by the experiments already made before the suggestion can be accepted. But you can all see by the mere mention of such a subject how intensely interesting the pursuing of these investigations further must be, and I believe Maxwell is at present making a measurement of this kind on a different plan from any that have been yet made. I have now spoken too long, or I should have described something of the methods already followed in this department, but they are already fully published, and can easily be referred to.

Now with respect to accurate measurement—theory was left far behind by practice, and I need not to be reminded by the presence of our

President how very much more accurate were the measurements of resistance in the practical telegraphy of Dr. Werner Siemens and his brother, our President, than in any laboratory of theoretical science. Whilst in the laboratory of theoretical science it had not been discovered that the conductivity of different specimens of copper differed at all, in practical telegraphy workshops they had been found to differ by from thirty to forty per cent. When differences amounting to so much were overlooked—when their very existence was not known to scientific electricians — the great founders of accurate measurement in telegraphy were establishing standards of electric resistance accurate to one-tenth per cent. Dr. Werner Siemens and our President were among the first to give accurate standards of resistance, and the very first to give an accurate system of units founded upon those standards. The Siemens unit is still well known, and many of the most important measurements in connection with submarine cables are stated in terms of that unit. By a coincidence, which

in one respect is a happy one, although there is
something to be said on the other side, the unit
adopted by Messrs. Siemens, founded on the
measurement of a certain column of mercury—
the Siemens unit produced and reproduced in
their accurate resistance coils—approaches somewhat nearly to the unit which in Weber's system
would be 10^9 or a thousand million centimetres
per second. This is so far convenient that
measurements in Siemens units are very easily
reduced to the absolute measure. The committee
of the British Association, of which our President
was one and I also had the honour to be a
member, proposed a method of measurement
which was carried out chiefly by Professors Clerk
Maxwell, Balfour Stewart and Fleeming Jenkin,
who laid down what is called the British Association unit of resistance, to which the name of
" Ohm," according to the advice of Mr. Latimer
Clark, was given in commemoration of one of
the great founders of electro-magnetic science.
Ohm being the man who gave us first the law
of currents in connection with electro-motive force

it was considered appropriate that his name should be given to this electric unit.

I may mention as a matter of great importance and interest in physical science, that a revision of the measurement of the British Association unit is being undertaken. An endeavour is now being made to measure with the greatest possible accuracy what is the value of the " Ohm " in terms of the absolute scale of centimetres per second. It will certainly come within a small percentage of being exactly ten thousand kilometres per second. One per cent. away from that amount, it may be: but that it may be two or three per cent. or four per cent. or one-third per cent. is of course possible ; as any one may judge by looking at the difficulties which will have to be met with in making the experiments. I will just say in connection with this electric measurement of the Ohm, that it touches on another point of measurement, that of heat. Joule in a quite independent set of experiments which I can only name, showed another way of arriving at similar results, and Joule's thermoelectric experiments taken in connection with other experiments of his on the dynamical equivalent of heat, show some disagreement from the British Association measurement of their unit of resistance. There is something to be reconciled here. Joule on the one side holds that the British Association unit, the Ohm, is too little, but on the other side, in Germany, Kohlrausch holds the Ohm to be a little on the other side of the exact thousand million centimetres per second. I believe if you eliminate doubt by the method of averages, Kohlrausch and Joule's experiments would show the British Association to be very nearly right, but I do not approve of that method of removing doubts, and we shall not be satisfied until both Joule and Kohlrausch are satisfied.[1]

I will now mention a number of experiments with electrometers which however, I am afraid, are of little interest to any one in the world, but myself. Here is the first attempt at a quadrant electrometer. It is well known now to many electricians, and a descriptive pamphlet regarding it

[1] See pp. 133—136 above.

has been issued. I really do not know, considering that the British Association report on electrometers has been republished in connection with the whole series of their Reports on Electrical Standards,[1] that I need go into detail with respect to any of these instruments. This is the very first portable electrometer, and I will tell you how it came into existence. I had one that I was very proud of, I am ashamed to say, in these days. I was proud of its smallness, and how easily it could be carried up to the top of Goatfell and back; there was one before that, the highest character of which was, that it was not heavier than a rifle. That was in the days of what Lord Palmerston called the "rifle fever," and I was touched a little with it at the time, being a rifle volunteer. I found that my electrometer weighed a pound less than my weapon. It only weighed thirteen lbs., and the rifle weighed fourteen lbs. I had that at Aberdeen, but it is not now to be found, although it has been searched for, or it would have been exhibited. Part of it

[1] E. and F. N. Spon, London, 1873.

the stand that was on the top of it, is shown. The next that followed was this one (Fig. 55). I got down the weight to about one-half, and I was perfectly satisfied then, and this one has gone up Goatfell a great many times; but it is fully described in my book,[1] and in the Report I have referred to I was showing it with great pride on one occasion to Professor Tait, and I said to him : " You should get one like that." He said, " I will wait until you can get one that you can put into your pocket. Get one the size of an orange, and then I will have it." That literally was the origin of this latest portable electrometer. I felt rather challenged by what he said, and in the course of my next run up to Glasgow, Mr. White, who is so indefatigable in making new things, and who has so admirable an inventive capacity, helped me in my endeavour, and we had something like this one. In the course of a month, this very electrometer (Fig. 56) was got into action. This is the first attracted disc electrometer. It differs from the portable electrometer

Papers on Electrostatics and Magnetism ; Macmillan, London, 1884.

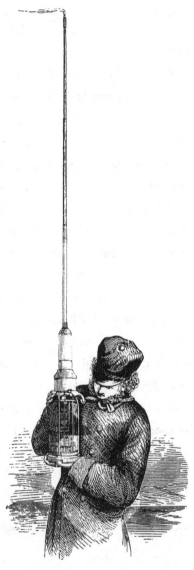

Fig. 55.—First Portable Electrometer.

as now known, merely in some minor details; the
moveable disc turns round with a micrometer
screw instead of moving up and down in a slide

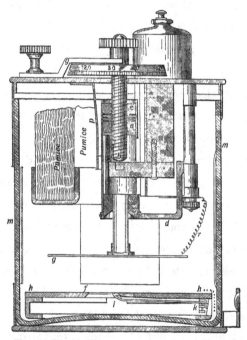

Fig. 56.--Portable Electrometer, present form : ⅔ full size.

In all other respects it is the same, except the
awkward arrangement for placing the pumice,
which with my great care, did not lead to any

accident, but with almost any other person would have led to the instrument being destroyed by the sulphuric acid getting shaken down into the instrument below. A more convenient arrange ment of the pumice is now made, but that is the only alteration besides the mechanical arrangement of the disc which is better in the portable electrometer as it now exists. Two of these instruments have been sent out with the Arctic expedition (of 1875-76)

Just one word of practical advice with respect to the electrometers. I have been continually asked how to keep them in order, and have fre quently heard complaints that these will not hold ; that they do not retain their charge. In each of these electrometers there is a glass Leyden jar, the heterostatic system being adopted in each of them. It is necessary the insulation should be very perfect, and then it all depends afterwards on the cleanness and dryness of the surface of the glass. If you will allow me to use the definition of Lord Palmerston, when he said that, "Dirt is matter in its wrong place," and to consider that water, or

any moisture on the inner surface of the glass
which ought to be perfectly dry—is "matter
in its wrong place," and is, therefore, dirt, you
will understand what I mean. If there is no dirt
on the glass it is certain to insulate well. But then
how to get the glass perfectly clean? In the first
place, wash it well with soap and water. If you
like you may try nitric acid, and then pure water,
or you may wash it with alcohol, and then with
pure water. I have gone through almost in-
cantations to get perfect cleanness of the surface of
the glass, but I doubt much whether I ever got
any result which I could not have got with soap
and water, and then running pure water over the
surface of the glass after it is done. Wash it
well, somehow or other. You may use acids, or
alcohol, if you like ; but I think you will generally
find that soap and water, and enough of clean
water to end with, will answer as well as anything.
Then shake it well, and get it well dry, but do
not use a duster, however clean, to dry it. Shake
the moisture off, and take a little piece of blotting
paper, and suck up very carefully any little portion

of water which may remain by cohesion, but do not rub it with anything that can leave shreds or fibres ; that is dirt. The finest cambric will leave on the glass what will answer Lord Palmerston's definition. When you have got the glass clean of everything except water, then dry it, and you will be sure to find it answer. The way to dry it, and to keep it dry, is to have the sulphuric acid in the proper receptacle. Each of these instruments has a receptacle for sulphuric acid, which must be freed from volatile vapours by a proper process ; boiling with sulphate of ammonia suffices. The sulphuric acid need not be chemically pure, but it must be purified from volatile vapours, and it must be very strong. I believe, oftener than from any other cause, these instruments fail to hold well because the sulphuric acid is not strong enough, and frequently, when an electrometer has failed, by putting in stronger acid the defect has been perfectly remedied.

INDEX.

INDEX.

RICHARD CLAY AND SONS, LIMITED, LONDON AND LUNGAY.

Printed in the United States
By Bookmasters